OTHER TITLES OF INTEREST FROM ST. LUCIE PRESS

Principles of Sustainable Development

Development, Environment, and Global Dysfunction: Toward Sustainable Recovery

Resolving Environmental Conflict: Towards Sustainable Community Development

Environmental Management Tools on the Internet

Economic Theory for Environmentalists

Sustainable Forestry: Philosophy, Science and Education

Ecology and the Biosphere

For more information about these titles call, fax or write:

St. Lucie Press
100 E. Linton Blvd., Suite 403B
Delray Beach, FL 33483
TEL (561) 274-9906 • FAX (561) 274-9927

S_L^t

ENVIRONMENTAL DESIGN PARTNERS, INC.
615 N. VICTORIA PARK RD.
FT. LAUDERDALE, FL 33304
USA

Sustainable Community DEVELOPMENT

PRINCIPLES AND CONCEPTS

Sustainable Community
DEVELOPMENT

PRINCIPLES AND CONCEPTS

CHRIS MASER

St. Lucie Press
Delray Beach, Florida

Phone: (561) 274-9906
Fax: (561) 274-9927

S$_L^t$

Published by
St. Lucie Press
100 E. Linton Blvd., Suite 403B
Delray Beach, FL 33483

CONTENTS

PREFACE

I have been struggling with the notion of sustainability and sustainable community development for many years, but it was only during my recent trip to Malaysia that enough of the pieces came together for me to consider a synthesis of what I think I understand. This synthesis, *Sustainable Community Development*, is possible only because it is an outgrowth of my earlier book, *Resolving Environmental Conflict*, published in 1996. *Resolving Environmental Conflict*, subtitled "Towards Sustainable Community Development," contains two chapters (10 and 11) that deal with this subject, both of which are co-authored by Christine Kirk.

Christine, a student in the University of Oregon's Department of Planning, Public Policy, and Management in 1992, was in a class to which I gave a guest lecture on "The Nature of Sustainability." Some months later, she sent me a copy of her senior thesis ("Community Development: The Inlet for Sustainable Change") with the following comment: "I very much enjoyed your style of presentation and your perception of sustainable development. Enclosed is my senior thesis. I would like you to read it because I believe that our perceptions of sustainable development are similar." Christine was correct.

I have taken as a partial skeleton for this book the two chapters on sustainable community development that Christine and I co-authored. While I have used some pieces of the earlier work more or less in tact, most have been rearranged, added to, deleted from, reconsidered, and so on. Nevertheless, as you read this book, please keep in mind that somewhere in the world is a very bright young woman named Christine Kirk, whose ideas influenced mine and are therefore a part of the current book.

If perchance you have read *Resolving Environmental Conflict*, you will find some overlap between that book and this one because the act of resolving environmental conflict is part and parcel of sustainable community development, and vice versa. Moreover, life and learning is a continuum along which I can discern no discrete entities. I have done my best, however, to minimize the overlap by assuming in this book that any existing conflicts have already been resolved and that a shared vision of the future toward which to build is already in place. As such, sustainable community development is the enactment of that vision. (When I sent the manuscript of this book out for review prior to submitting it to be published, it was clear that those reviewers who were familiar with *Resolving Environmental Conflict* had an easier time grasping fully the principles and concepts explored in this book.)

Sustainable community development is important because civilizations have been marked by their birth, maturation, and demise, the latter often brought about by decay of social morality and by uncontrolled population growth that outstripped the source of available energy. In the past, survivors could move on to less populated, more fertile areas as their civilizations collapsed. Today, there is nowhere on Earth left to go!

Yet, by denying the lessons of history, our civilization is repeating the mistakes of history—but this time on a global scale—by destroying the very environment from which it sprang and on which it relies for continuance. It thus seems that humanity is not much daunted by the prospect of misery—as long as it is misery that is delayed or will happen to someone else. Be that as it may, we cannot avoid social/environmental catastrophe if, as Lester Brown says, "we keep sleepwalking through history."

If our civilization is to survive, we must *begin now* to learn history's most consistent and challenging lesson: to save our civilization, we must conquer ourselves. In the material world, self-conquest means bringing one's thoughts and behaviors in line with the immutable physical/biological laws governing the world in which we live, such as the law of cause and effect and the law of entropy. In the spiritual realm, this means disciplining our thoughts and behaviors in accordance with the highest spiritual/social truths handed down through the ages, such as treat others as you want them to treat you.

The outcome of self-conquest is social/environmental harmony (manifested through social/environmental sustainability), which must be the next cultural stage toward which we struggle. Social/environmental harmony is that stellar space beyond self-centeredness for which there needs to be an ethical foundation and a world government.

Many people with whom I have conversed over the years have lent, at best, a cynical ear to these ideas. In fact, a recent review* of one of my books (*Sustainable Forestry: Philosophy, Science, and Economics*) is a case in point:

> In his conclusion, Maser advocates increased environmental protections for U.S. forests to demonstrate to the world our commitment to sustainability. Say what? U.S. consumers are the most voracious in the world and are hardly candidates for global role models. In fact, one could argue that our current environmental protections, coupled with free trade, simply export the production of "messy" goods to other parts of the globe. Such "exporting" of environmental degradation from developed nations shares some responsibility for the appalling environmental conditions found in many developing countries.

Implicit in this statement is the often-heard refrain that this is simply the way we are, as though we somehow cannot change. I submit, however, that it is precisely *because* we are the most voracious consumers in the world that *we must* become the world's role models if sustainability is to be anything more than a hopeful daydream. Environmental protection is the *necessity to which economics must adapt*—not the other way around. Economics without humility is every bit as dangerous as science without morality. And Western society (particularly in the United States) is at best unhumble or we would not be forcing the rest of the world to fulfill our economic appetites, which are the outgrowths of our appalling lack of self-discipline.

True progress toward an ecologically sound environment and a socially just culture will be initially expensive in both money and

---

* Rob Lilieholm. 1995. *American Scientist* 83:578–580.

effort, but in the end will determine whether our society survives. The longer we wait, however, the more disastrous becomes the environmental condition and the more expensive and difficult become the necessary social changes.

No biological shortcuts, technological quick fixes, or political promises can mend what is broken. Dramatic, fundamental personal and social change is necessary. It is not a question of can we or can't we change, but one of will we or won't we change. Change is a choice, a choice of individuals reflected in the collective of society and mirrored in the landscape.

If we fail in our bid for social/environmental sustainability, our civilization will follow at least 26 other great civilizations down the corridors of extinction because, as British historian Arnold Toynbee concluded, they could not or would not change their direction, their way of thinking, to meet the changing conditions of life.

"We cannot say [what will happen] since we cannot foretell the future," says Toynbee. "We can only see that something which has actually happened once, in another episode of history, must at least be one of the possibilities that lie ahead of us."* And social collapse through moral decay, uncontrolled population growth, and unrestrained economic exploitation that outstrips the source of available energy has been disturbingly repetitive throughout the annals of history.

Consider, for example, that circumstantial changes in a community's resource base may require a dramatic shift in its frame of reference, its identity, such as a community built around lumbering when harvestable timber runs out or a coastal town built around commercial fishing when the fish stocks are depleted. A community's challenge thus becomes how to change from being narrowly specialized, and therefore dependent on a given resource for its cultural identity and economic survival, to becoming diversified in a sustainable fashion, ecologically, economically, and culturally. For a democratic community to be sustainable, it must be active with intelligence, which means the community must take the time to reflect on the meaning, purpose, and direction of its activity if social/environmental sustainability is to prevail.

---

* Arnold Toynbee. 1958. *Civilization on Trial and the World and the West.* Meridian Books, New York, 348 pp.

Instead of focusing on the fear of loss or fear of change by ruminating in the past, a community must shift its focus and use change as a fulcrum of hope and choice for a more sustainable direction in the present for the future, which brings us to the notion of sustainable community development. I have seen various attempts to define "sustainable," "development," "community," "sustainable development," and "sustainable community development." In fact, I have attempted a few definitions myself and have come to the conclusion that a definition of sustainability, as my friend Duncan Taylor says, "is elusive, like a horizon, receding whenever one attempts to discover it boundaries."

I am now convinced that none of these terms, either singly or in combination, is completely definable linguistically when discussing sustainable community development. Rather, they compose the pieces of a dynamic, interactive, interconnected, interdependent system of being, which by its very nature is definable not in words but rather by the functional interactions of its pieces as a whole.

Community, in the sense of sustainable development, focuses on the primacy and quality of relationships among the people sharing a particular place and between the people and their environment, particularly their immediate environment.

In the sense of sustainable community, *development* means personal and social transformation into a higher level of consciousness of cause and effect and a greater responsibility to be one another's keepers through all generations. Development does not mean continual physical/economic growth, which is neither biologically nor economically possible without destroying the umbilicus between ecosystem and economy. (An ecosystem includes all living organisms interacting with their nonliving physical environment, considered as a unit.)

Sustainability, in the sense of community development, is the act of one generation saving options by passing them to the next generation, which saves options by passing them to the next generation, and so on. Sustainability will demand a shift in personal consciousness—from being self-centered to being other-centered.

While I do not pretend to have the answers, and while my perception of the truth may in fact be incorrect in some cases, I may nevertheless be able to help frame some of the questions. Therefore, I intend to explore various aspects of sustainability (within the context of community development and its relationship to landscape), recognizing that sustainability is a continual process, instead of some finite end point at which one arrives.

Although I use illustrative examples from a number of places in the world in my discussion, I naturally keep coming back to those I know best—the local examples of my hometown and its surroundings in western Oregon. And here an important point needs to be made.

It has become clear to me over the years that many, if not most, people in Western society tend to focus on the differences among things rather than their commonalities. Yet it is the commonalties of the unifying principles and concepts that keep the differences within the perspective of a functional whole. So I ask you to focus on the principles and concepts offered in this book; to the extent they are correct, you will find that the specifics largely fall into place.

ACKNOWLEDGMENTS

To me, one of the greatest pleasures of writing a book is saying "thank you" to the people who have given generously of their time to help me rise toward my ever-elusive goal of excellence. I am deeply grateful for that help.

I have for many years struggled with the notion of sustainability and sustainable community development. But it was in Malaysia that I experienced the integration of a people in a way that I have never before encountered in all my travels. The Malaysian people showed me that real integration *is* possible, and through my experience in Malaysia, this book finally came together.

I extend my sincere appreciation for the warm hospitality offered me by the Malaysian people; to Dr. M.N Salleh, former director of the Forestry Research Institute of Malaysia (FRIM), who invited me to Malaysia and made the trip possible; and to the staff of FRIM, who helped make my travel arrangements and saw to it that my stay was most pleasant and memorable.

The following people, in alphabetical order, were kind enough to read the entire manuscript, openly and frankly challenge my ideas, and make numerous improvements: Ruth Bascom (Mayor of Eugene, Oregon), Helen Berg (Mayor of Corvallis, Oregon), Paul N. Goodmonson, Jr. (Businessman and Financial Advisor, Corvallis, Oregon), Kevin Smith (Intergovernmental Affairs Manager, Oregon Economic Development Department, Salem, Oregon), and Robert F. Tarrant (Professor Emeritus of Forest Science, Oregon State University, Corvallis).

Sandy Pearlman did the marvelous job of copy-editing and manuscript preparation that I have come to expect from her. Working with Sandy is for me a privilege.

To my wife, Zane, I extend a special "thank you" not only for proofreading the galley but also for once again accepting graciously the many hours, both day and night, that I spent working on this book.

ABOUT THE AUTHOR

C hris Maser spent over 20 years as a research scientist in natural history and ecology in forest, shrub steppe, subarctic, desert, and coastal settings. Trained primarily as a vertebrate zoologist, he was a research mammalogist in Nubia, Egypt (1963–1964) with the Yale University Peabody Museum Prehistoric Expedition and was a research mammalogist in Nepal (1966–1967) for the U.S. Naval Medical Research Unit #3 based in Cairo, Egypt, where he participated in a study of tick-borne diseases. He conducted a three-year (1970–1973) ecological survey of the Oregon coast for the University of Puget Sound, Tacoma, Washington. He was a research ecologist with the U.S. Department of the Interior Bureau of Land Management for twelve years (1975–1987)—the last eight studying old-growth forests in western Oregon—and a landscape ecologist with the Environmental Protection Agency for one year (1990–1991).

Today he is an independent author as well as an international lecturer and facilitator in resolving environmental disputes, vision statements, and sustainable community development. He is also an international consultant in forest ecology and sustainable forestry practices. He has written over 240 publications, including his last five books: *Forest Primeval: The Natural History of an Ancient Forest* (1989, listed in School Library Journal as best science and technical book of 1989), *Global Imperative: Harmonizing Culture and Nature* (1992), *Sustainable Forestry: Philosophy, Science, and Economics* (1994), *From the Forest to the Sea: The Ecology of Wood in Streams, Rivers, Estuaries, and Oceans* (1994, with James R. Sedell), and *Resolving Environmental Conflict: Towards Sustainable Community Development* (1996). Although he has worked in Canada, Egypt, France, Germany, Japan, Malaysia, Nepal, Slovakia, and Switzerland, he calls Corvallis, Oregon, home.

To the children of the world.
I am deeply sorry for the way
we adults have by and large treated you.
May God bless you—and forgive us.

Man is much more afraid of the Light than he is of the Dark and will always shield his eyes against a truth which is brought to him prematurely. He will throw stones at it or even crucify it in order to remain in the comfortable shadow of his ignorance. But that is human nature and Man must not be condemned for his unconsciousness.

Alan Oken
Astrologer

Any man who has the brains to think and the nerve to act for the benefit of the people of the country is considered a radical by those who are content with stagnation and willing to endure disaster.

William Randolph Hearst
American newspaper publisher

WHY BE CONCERNED WITH SUSTAINABILITY?

n the early days, when travel was limited to horse and buggy, communities were in some ways more diverse and convenient than they are today because basic necessities, such as food markets, general stores, doctors' offices, and post offices, were within a short distance of one another. With the advent of the automobile, however, things could be spread out more, but generally were still within convenient distances of one another. When I was a boy in the early 1940s, for example, my family lived about four miles out of town. But once we drove into town and parked the car, we could do virtually all of our shopping within walking distance of where we parked.

The community in which I grew up was friendly. No one I knew of locked either their home or their car. I'm not even sure that in the early years my parents knew where the key to the house was. The neighbors all visited with one another and people were friendly. My father knew the man at the gas station, which in those days was a "service station," the president of the bank, and went to the same barber for years. My mother knew the milkman and the postman, both of whom always had time for a kindly word. My mother also knew the druggist and the grocery clerk. And our family doctor made house calls with a smile. In short, people cared, genuinely cared, about one another's welfare. People seemed to be important to one another in those days, and there was a corresponding sense of mutual support.

Even the landscape around my home town was friendly. The town was surrounded by fields and forest, which were connected by swift forest streams that fed meandering valley rivers. And I was free to wander over hill and dale without running into a "no trespassing" sign on every gate and seemingly every other fence post.

The code of the day was to leave open any gate that was open and to close after one's passage any gate that was closed. It was also understood that one was free to cross a farmer's property as long as one respected the property by walking around planted fields rather than through them. And, if I asked permission, I could wander, hunt, fish, and trap almost anywhere I wished.

Much of the Coast Range and most of the Cascade Range of Oregon that I knew as a boy were covered with unbroken ancient forest and clear, cold streams from which it was safe to drink. Although the streams were still filled with trout and salmon, the forests and mountain meadows were already devoid of wolf and grizzly bear.

In the valley that embraced my home town, the farmers' fields were small and friendly, surrounded by fencerows sporting shrubs and trees, including apples and pears that proffered delicious fruit, each in its season. In spring, summer, and autumn, the fencerows were alive with the colors of flowers and butterflies and the songs of birds. They harbored woodrats and rabbits, pheasants and deer, squirrels and red valley foxes. The air was clean, the sunshine bright and safe, and the drinking water amongst the sweetest and purest in the world.

When World War II came along, the seeds of change were sown with respect to community. The war effort pushed mass production to new levels and brought impersonalization of humans killing humans to the fore with such labeling of cartons containing weapons as "mine, one, anti-personnel," which indicated that the person the weapon was meant to kill was simply a military abstraction.

Although World War II eventually drew to a close, the impersonalization of mass production carried over into the post-war boom years. Gone was the simple wisdom of building communities and neighborhoods within communities for people; it was replaced by the strategies of massive wartime production developed in defense factories.[1]

Instead of paying careful attention to all the necessary details that made a neighborhood within a community a pleasant, nurturing place to live, the post-war planners focused narrowly on churning out great numbers of cracker-box houses on an assembly-line scale. Although they succeeded in cranking out modern housing for millions of middle-class citizens, they sacrificed quality, neighborhood, and community in favor of quantity.[1]

Little or no thought was given to how children might get from home to school to basketball practice or ballet lessons and home again. Little or no thought was given to what would happen when suburbanites grew old and could no longer drive an automobile. At the same time, however, the government offered incredibly generous loans to returning GIs so they could buy new houses. But there was no money available to purchase and fix up existing houses, and so communities began their rapid decline.[1]

Towns, including mine, started to sprawl rapidly in largely un-planned ways. Cookie-cutter houses were concentrated in develop-ments that were isolated from everything else dealing with community.

Speed rather than care began creeping into the building trade, and I watched as houses sprang up in blocks and lines and circles, built for speculation. As speculation crept into the housing market, speed, sameness, and clustering became marks of efficiency and greater profit, setting the tone for the future—a tone reflected in the night sky as the once brilliant stars of the Milky Way disappeared into a seemingly eternal mask of light pollution.

With the stage set by the post-war housing industry, things began to change noticeably as corporate depersonalization commenced its insidious cancer-like growth into the heart of community. Shopping malls were connected by roads that became bigger, straighter, faster, and increasingly went through prime agricultural land. Then came larger and larger subdivisions with cheaper and cheaper ticky-tacky tract housing, some of which was constructed in floodplains or on unstable soils. I remember, for example, looking at such a house in Las Vegas, Nevada, in 1990, where the kitchen counters were literally pulling away from the wall. I asked the builder if the situation was going to be corrected. "Nah," he said, "somebody will buy it."

Centralization had arrived on the landscape as it had earlier in corporations. Driving on superhighways became a necessity, and with it came pollution of air and water, which increased with every extra mile that had to be driven and every additional automobile on the road. And the gentle motion and the relaxed pace of the traditional street gave way to ever-increasing speed. As author Jean Chesneaux observed: "The street as an art of life is disappearing in favour of traffic arteries. People drive through them on the way to somewhere else." There is no word in English with a positive connotation for going slowly or lingering on streets as a way to participate in a sense of community.

People started losing their sense of connection as centralization within urban sprawl increasingly specialized the human landscape, and communities began falling apart. A sense of place—of a familiar, friendly community, where everyone left their homes unlocked—gave way to a sense of location as more and more people became transients. Many were and are moved about at corporate or agency will. (Las Vegas, Nevada, had such a transient population in the two years I lived there that the phone company printed a huge, entirely new phone book every six months.)

By the time I was a teenager, it had become necessary to lock the doors to our house, and "no trespassing" signs proliferated across the landscape. A sense of distrust had begun its insidious invasion throughout the once closely knit human bonds of mutual caring that in days gone by had characterized a community.

Outside of town, the forests were being cut at an exponential rate, including the town's water catchment, endangering such species as the northern spotted owl and marbeled murrelet. The forested streams, where as a lad I drank of their sweet water and caught native cutthroat trout, now have waters unsafe to drink. Clearcut hillsides began eroding as forests were converted to economic tree farms. Gone are most of the great native trout and the wild salmon that graced the streams from which I drank. Gone are the great flocks of bandtailed pigeons that once greeted me in forest and fen. Gone are the elk and bear that I used to see within ten miles of my house. Gone is the forest of centuries. In its place are acres of comparatively lifeless economic tree farms, some of which may live but a little longer than I.

At the same time, I watched helplessly as the small protected fields of the personable family farms increasingly gave way to larger and larger naked, homogeneous fields of corporate-style farms as fencerows were cleared to maximize the amount of tillable soil, to squeeze the last penny from every field. With the loss of habitat along each fencerow, the bird song of the valley was diminished in like measure, as was the habitat for other creatures wild and free.

Gone are the fencerows with their rich, fallow strips of grasses and herbs, of shrubs and trees, which interlaced the valley in such beautiful patterns of flower and leaf with the changing seasons. Gone are the burrowing owls from the quiet secluded fields that I once knew. Gone is the liquid melody of the meadowlark that I so often heard as a boy. Gone is the fencerow trill of the towhee. Gone are the song sparrows, Bewick's wrens, yellow warblers, and MacGillivary's warblers. Gone are the woodrat nests, the squirrels, and the rabbits. Although these species may still occur along the edge of the valley and in isolated patches of habitat, they are gone with the fencerows from the agricultural fields of the valley floor.

Today, compared with the time of my youth, the valley's floor offers little in the way of habitat, other than a great depersonalized open expanse of silent, naked fields in winter and a monotonous sameness under the sun of summer. And then came the coup de grâce—zoning!

Zoning is the predetermined division of land based on what people think its best potential use is. While "best potential use" is meant to be in the most sustainable social/environmental sense, it is usually parlayed into disguised economic growth through political power from which a few benefit financially. Zoning, therefore, like most ideas that are used as tools in human endeavors, has both beneficial and detrimental effects with respect to a sense of community within a landscape. Which of these alternatives comes to the fore depends, of course, on the self-centeredness or other-centeredness (the personal morals and integrity) of those who make the decisions.

With respect to a community itself, zoning has been largely detrimental in the sense that it has been used to foster such things as concentrated shopping malls and isolated housing developments, often in irreplaceable agricultural and/or forested land. (See Gary Gardner, "Preserving Agricultural Resources," for a synopsis of what is happening to agricultural resources.[2]) Such centralized specialization fragments the ability of people to meet their daily requirements within a convenient area. Although such convenience may not, on the surface, seem of much importance, it is in at least two ways.

In all the small communities in which I have lived, both in the United States and abroad, the people could fulfill their basic daily requirements within walking distance of their homes, which meant getting exercise and keeping the air cleaner by leaving their cars behind. This convenience, this *centralized generalization* of goods and services, also allowed people to meet one another on the street and in the shops, where they got to know one another and often stopped to exchange pleasantries. There were also places, such as a particular cafe or soda fountain, a particular shade tree, or a community well, where local people would gather daily and visit.

As people got to know one another, they became familiar with what each person did and whom they could rely on when in need of help. Because people knew one another, they looked out for one another in a free exchange of mutual caring. Of course there were social problems and interpersonal conflicts, but there was the greater incentive of mutual well-being to work them out.

With the advent of fragmentation through *centralized specialization*, people were (and are) forced increasingly to drive from their homes to go to the grocery store, the doctor, the post office, the hair stylist, and so on. This forced reliance on an automobile not only isolates people from the opportunity of getting to know one another by walking familiar streets together but also isolates them in the shopping malls, where people are in essence visitors from foreign housing developments. And because such centralized specialization necessitates driving automobiles, it dramatically increases the problem of pollution, both air and water.

Gone are most of the friendly communities of old, those centered around daily living within walking distance, where open spaces from the surrounding landscape were integrated within the community itself. Increasingly, today's communities are what I would call "suburban block communities," where a few neighbors may get to know one another, at times almost as a defense against the larger world.

I also find that people tend to defend these small enclaves, often like special interest groups defending an intellectual position against an opposing point of view. Other people cluster within fortified "communities" surrounded by high walls and locked gates (at times with guards) to "protect" themselves from their neighbors on the outside. I see this as a sign of increasing isolation, which fosters neighborhood isolationism in the form of social fragmentation instead of overall communality.

This attitude should come as no surprise, however, because today's real communities, those of long existence and still somewhat closely knit, are increasingly being held hostage by the economy of super-highways, shopping malls, and corporate chain stores, which import goods from outside of the community despite the availability of local supplies. All of this fits within an interesting article in which Jon Margolis discusses the ideas of a self-taught unconventional Italian philosopher, Giovanni Battista Vico, who was born in 1668.[3,4]

Giovanni Battista Vico had a great deal to say about the 1990s, says Margolis, even though he completed his major work (*Principles of a New Science Concerning the Common Nature of Nations*) in 1725 and died in 1744. In his great work he propounded a cyclical theory of history in which human societies progress through a series of stages from barbarism to civilization and then a return to barbarism.

In the first stage (called the Age of the Gods), religion, family, and other basic institutions emerge. Then comes the Age of Heroes in which the common people are subjugated by a dominant class of nobles. In the final stage (the Age of Men), the people rebel and win equality, but in the process, society begins to disintegrate. "The nature of peoples," said Vico, "is first crude, then severe, then benign, then delicate, finally dissolute."[3,4]

In short, as people make life for themselves increasingly better materially, they descend into spiritual, moral, and intellectual decay, which manifests itself in the corruption of justice and governmental processes. "Men first feel necessity," Vico wrote, "then [they] look for utility, next attend to comfort, still later amuse themselves with plea-sure, thence grow dissolute in luxury, and finally go mad and waste their substance [which can be interpreted as the natural resources of their mutual environment]."[4]

The end of the process is apparent, according to Vico, when "each man [is] thinking only of his own private interests." Then arises the next stage "from an excess of reflection or from the predominance of technology," as stated in the *Encyclopedia Britannica* article on Vico. When materialism finally degenerates into a systematic examination of itself, people are "made inhuman by...reflection." Thus, said Vico, human society progresses from forests to huts to villages to cities, "finally [ending at] the academies."[3]

"Sounds familiar, doesn't it?" asks Margolis. "In the 370 years since the Pilgrims converted some of the forest to huts and then villages, Americans have gone through the entire process." We are now domi-nated by people who think only of their own private interests at the expense of both the environment and the generations of the future, by people reflecting on how other people think only of their own private interests, and by people who do both simultaneously.[3]

There are, however, people trying to reverse this historical trend through the practice of conscious (=voluntary) simplicity in an effort to get back to community—back to the future as it were. Reclaiming community is critical in these days of whirlwind change and massive international economies that encircle the globe like entangled webs of so many competing spiders. Community brings a concreteness, a common frame of reference, into the lives of ordinary people who increasingly see the world as an unsafe maze of unrelated abstractions.

In this sense, zoning, which helped to disintegrate the local community I knew as a boy, can be used to help reclaim a sense of community by recreating the basic physical structure necessary for community. Such zoning, while protecting the centralized diversity around which communities flourish, will fly in the face of those major corporations that use communities merely as colonies of profit. Although zoning is not meant to eliminate corporations, it is intended to protect communities from the competitive tentacles and centralizing control of national and international corporations.

But simple zoning is not enough to reclaim community. To be an effective tool of sustainable community development, zoning must be done carefully, thoughtfully, and with the future in mind while making current decisions. This will require an understanding and, where appropriate, acceptance and application of things discussed in this book.

To reclaim community does not mean building semi-self-contained gated communities, as is the current trend in many places. Reclaiming community means that we must relearn much of what we have forgotten. We must remember and reconstruct the cultural values we once had in community but lost sight of. People must harbor such a deep longing for community that they are totally committed to practicing community, which is based first and foremost on meaningful interpersonal relationships. Only then will recreating a community's physical attributes have any significance.

Reclaiming the physical attributes of community will mean once again designing communities around centralized hubs of generalized variety, such as small, locally owned stores (which could be designed with living quarters above them, as in Europe) that serve a local area without forcing people to drive their automobiles. It will mean getting back to designing streets that are friendly to people and building houses with front porches. Alleys between houses will have to be reinstituted to concentrate garages, automobiles, and garbage cans behind the houses so people can more freely interact in a more beautiful setting. Placing the garages in the back alleys will leave space for gardens alongside and in front of houses so people can once again walk the streets, admiring flowers and visiting with their neighbors.

In addition, neighborhoods would have mixed housing so that they do not become isolated enclaves of rich versus poor, black versus white, and so on. If we truly want community, we must accept the richness of our ethnic diversity by focusing on our unity as Americans, a lesson we could well learn from the Malaysians (which will be discussed later). If we do not learn to focus on our unity, we will continue to focus on our differences, which translates into racial seg-regation—not on paper perhaps, but in the streets, where racism cannot be hidden.

And there would be an opportunity to reintroduce open spaces into a real community setting, instead of just a consolidated acre or two given grudgingly by a developer interested only in the financial bottom line. Having said this, I realize that not all developers are so stingy, but the vast majority with whom I have dealt over the years have been, which brings me to zoning as a way of reintegrating community and landscape to capture the synergism of ecology and culture.

For example, an available supply of good drinking water is all-important to the sustainability of a community and to the protection of the value of its real estate. It would be wise, therefore, to zone areas from which community water comes in such a way that they are protected from the possibility of chemicals getting into the soil and contaminating the water, be they from agriculture, forestry, or industry. Such zoning would also benefit critical water catchments already purchased as open space to protect the infiltration and storage of water in the ground.

Will such wisdom in zoning happen while there is still time? I don't know. I sat for three years on the Environmental Advisory Committee of our local county, and for three years I tried to get the city and county to address the issue of sustainable water for the next two to three centuries. I told them that if they continued to allow the con-struction of housing developments without a vision of the future that included protecting the available water, the wells would go dry. No one listened.

Within one year after my term on the committee expired, the wells started going dry, and still no one is doing anything to protect the water catchments around the town. Moreover, the quality of the water has already declined drastically. Gone is the pure, sweet water of my youth. Today it is brownish and tastes of pollution, a flavor that becomes dramatically worse if the water is allowed to remain in a glass overnight.

Communities that are surrounded by agricultural land and/or forested land fail to comprehend that they are in many ways blessed with wealth almost beyond compare—provided they protect those lands for the sustainability of agriculture, forestry, and the production of water. If these lands are converted to housing subdivisions, shopping malls, or roads, a few people will profit immediately, but the majority will pay an increasingly stiff price over time as the community becomes less and less sustainable ecologically, culturally, and economically for all the reasons outlined in the following pages.

In addition, agricultural lands and forested lands are finite in number of acres, especially prime acres. Once lost to nonsustainable development, they are gone. And the effects are all but irreversible. So, once again, zoning for specific sustainable uses of land that are in keeping with sustainable community development will accrue incalculable value for the community wise enough to zone wisely.

Riparian areas and floodplains, which will be discussed later, are also essential to zoning in such a way that they are protected intact because of the vital ecological functions they perform. A community that does not protect these areas will eventually pay a very high price for its short-sightedness, both ecologically and economically. If you doubt this, ask the people living along the Missouri and Mississippi Rivers, or any number of other rivers for that matter.

In the end, we must ask ourselves what we really want: sustainable communities that afford quality lifestyles or a rapidly decreasing quality of life worldwide in which a very few people get rich while the majority get progressively poorer. Unfortunately, we are now headed for the latter, much as some people attempt to deny it. I cannot deny it, however, because I travel too much to put my head in the economic sand of corporate platitudes and empty political promises.

In Japan, for example, both the stages of life and education are compressed into a massive metaphorical G-force. This metaphorical G-force squeezes the childhood out of the child and provokes the highest suicide rate among children anywhere in the world.[5]

Could such a thing happen in the United States? Yes, it could. It has already begun. Here, however, the suicide is indirect; here, it is children and adolescents murdering one another. Today, in the once safe streets of my boyhood town lurk rapists and murderers. Now the streets are unsafe by night, particularly for women, and sometimes even by day.

"But," you may say, "trying to cure our social ills, let alone those of the world, seems so hopeless. Why even bother with the concept of sustainability?"

I cannot answer this question for you. I can only answer it for myself. I am concerned about sustainability because it means the ultimate difference between the survival *or* the extinction of human society, and perhaps of humanity itself. And I find in humanity a most wonderful experiment in relationship—that within ourselves spiritually; that with one another politically, present and future; and that with our environment. This trio of relationships, carried to its zenith, has, I believe, the potential to benefit the whole of the Universe. And because I have in my life felt a little of the joy and value of community in the form of relationships that foster a sense of mutual well-being and social/environmental harmony, I feel the human experiment is worth saving. So I do what I can.

SUSTAINABLE DEVELOPMENT: THE CONCEPT

Although our knowledge of the way Nature works has increased dramatically since our dawning, we still do not have the answer to one of our most pressing questions: Will the ecosystems of the future, which we are today shaping, continue to function in such a way that the quality of human life we have come to expect will continue? Although we shall never fully know the answer, we will continue to pursue the kinds of information that hint at the answer. In the meantime, the world continues changing in ways we do not expect and cannot predict, and we change our ideas, albeit slowly, as we experience the ever-changing present.

The only constant in the Universe is change, and change is the one process that we in Western society resist the most. Change is a continual process, be it an expanding Universe, an aging forest, or a shifting economy.

With this in mind, consider the term *issue*, which is a point of debate, discussion, or dispute; a matter of public concern; a culminating point leading to a decision. An issue, which becomes a focal point of public concern and debate, is based on a change in some circumstance.

This alteration in circumstance at first necessitates, and finally precipitates, a change in perception, which ultimately precipitates a change in action. It is in dealing with change that the challenge of the future lies today before society, a challenge we must accept, face, and deal with or see society as we know it perish.

An awareness of disastrous consequences, brought about by historically proven unwise choices, should encourage us to change the way we do things so that we may alter an outcome. But before we can alter the outcome of any historical trend, we must ask fundamentally different questions than heretofore have been asked. After all, an answer is only important when the right question is asked.

One such question might be: What is sustainable development? Sustainable development, as I perceive it, is a nonlinear process of systems thinking through which the social significance of nonmaterial wealth, qualitative values, and the heritage of both cultural diversity and identity can be accounted for in social decision making.

In this context, sustainability has at least ten essential elements: (1) understanding and accepting the inviolate physical principles governing Nature's dynamics; (2) understanding and accepting that we do not and cannot manage Nature; (3) understanding and accepting that we make an ecosystem more fragile when we alter it; (4) understanding and accepting that we must reinvest in living systems even as we reinvest in businesses; (5) understanding and accepting that only a unified systemic world view is a sustainable world view; (6) accepting our ignorance and trusting our intuition, while doubting our knowledge; (7) specifying what is to be sustained; (8) understanding and accepting that sustainability is a continual process, not a fixed end point; (9) understanding, accepting, and being accountable for intergenerational equity; and (10) understanding, accepting, and being accountable for ecological limitations to land ownership and the rights of private property.

In turn, the expression and viability of these ten elements depend on four human relationships and two questions. The relationships are: (1) *intra*personal, (2) *inter*personal, (3) between people and the environment, and (4) between people in the present and those of the future. The questions are: (1) when is enough enough and (2) are the consequences of our decisions reversible?

Development, in the context of sustainability, is a process of directed change, of social maturation if you will. Development as social maturation is a psychological transformation, which causes people to voluntarily reach beyond the immediacy of self-centeredness in a conscious effort to be accountable for the effects present decisions may cause in the future. Development thus becomes a process of change guided by the principles of social and environmental justice—for all living things, not just humans, present and future.

There is a move on the part of some American ecologists and people in both ecological restoration and conservation to want to restore ecosystems to conditions that existed prior to European settlement. I do not think this is necessary, wise, or even feasible. Instead, we must learn to maintain ecosystems in a biologically sustainable mode with our own human presence included. We must finally come to grips with the fact that we humans must pay our own way. The cost of our presence in an ecosystem must be accounted for in its sustainability.

SUSTAINABILITY

To achieve the balance of energy necessary to maintain the sustainability of ecosystems, we must focus our questions, both social and scientific, toward understanding the physical/biological governance of those systems. Then we must find the moral courage and political will to direct our personal and collective energy toward living within the constraints defined by ecosystem sustainability and not by political/economic desires.

Because the systems we are designing by our interaction with our environment are continually changing our environment (all of it, if in no other way than through air pollution), conditions prior to European settlement seem irrelevant. The systems we are creating are becoming ever further removed from the types of ecological balances that characterized pre-European conditions. Remember, however, that the first Americans were here long before the Europeans, and they had already altered the pre-human condition.

"...major civilizations have exploited resources and paid the price. Less grand cultural adventures [than ours], that have lasted longer, have been held in the vice grip of what nature can spare: the hunters and gatherers. We as a civilization find ourselves at a cultural watershed, where we cannot return to the existence of a noble savage, nor can we persist in the reckless activities of rapacious exploitation [driven by insatiable consumerism]."[6] Our challenge today is to mature suffi-ciently in personal and social consciousness to create and/or maintain sustainable ecosystems in the present for the future.

Be forewarned, however, that sustainability is an absolute. A system is either sustainable in a given state or it is not; there are no degrees of sustainability. Although sustainability is not a condition in which a compromise can be struck, the decisions leading toward that sustainability often require compromise. Seeking sustainability to a degree, which appears like an innocuous compromise, defeats sustainability altogether. Leave one process out of the equation or in some other way alter a feedback loop, and the system as a whole will gradually be deflected toward an outcome other than that which was originally intended. It is thus critical during any planning process to consider carefully the elements of sustainability and the relationships that make them work and to understand that they must be inviolate if the plan is to succeed.

The Elements

Although the following elements of sustainability are, to my mind, givens and not negotiable, I have not attempted to cover them all. Only those elements that I have found to be common impediments to sustainability in community settings are discussed.

First Element: Understanding and Accepting the Inviolate Physical Principles Governing Nature's Dynamics

Of the physical principles governing Nature's dynamic balance, three, as far as we know, are inviolate.[7] The first principle, *the law of con-servation of mass*, simply states that mass can neither be created nor destroyed; therefore, materials cannot really be "produced" or "con-sumed." The mass of a material remains the same while its form is altered from a raw material to finished products, wastes, and residuals without a change in quantity. Thus, over time, the amount of matter moving through a stable system must equal the amount stored in it plus the amount moving out of it.

Einstein's "Special Theory of Relativity" states that the energy pro-duced (E) is equal to the amount of equivalent mass converted in a transitory state as the mass of radiant (or other forms of) energy (m)

times the square of the speed of light (c, or 3×10^8 m/s) or $E = mc^2$. This conversion of mass energy to other forms of energy occurs in many processes; for example, it occurs in everything from the energy radiated by an ordinary light bulb, which is too small to measure in most cases, to the measurable mass energy radiated by our sun and other stars.

The second principle, *the law of conservation of energy*, states that energy can neither be created nor destroyed. Thus, while energy can be changed in form and distribution, the quantity remains the same. The notion of either "energy production" or "energy consumption" is therefore a non sequitur.

Energy is neither produced nor consumed; it is only converted from one form to another. For example, when the energy stored in fossil fuels is released ("consumed"), creating thermal, mechanical, or electrical energy, only the form of the energy has changed.

The third principle, *the second law of thermodynamics*, states that the amount of energy in forms available to do useful work can only diminish over time. The loss of available energy thus represents a diminishing capacity to maintain "order," which increases disorder or entropy. When considering the notion of social/environmental sustainability, it is vital to understand that an "expenditure" of energy means to convert a useful or available form of energy (that with which work can be done) to a less useful or less available form.

Some forms of energy conversion, however, have problems of pollution associated with them, such as the conversion of nonrenewable fossil fuels to electricity. In the process of converting coal to electricity, pollutants, such as sulfur dioxide, are spewed into the air and carried hundreds or thousands of miles from the coal-fired power plants that produced them in a phenomenon known as acid deposition, commonly called acid rain.

Second Element: Understanding and Accepting That We Do Not and Cannot Manage Nature

Sustainability is predicated on the humility to admit and accept that we humans do not and cannot manage Nature, that we are not in control. Rather, we treat Nature in some way and Nature responds to that treatment.

We, nevertheless, cling to the concept of management. In fact, the current buzzword "planetary management" has a nice ring to it, because it places the blame for things run amok on the planet and not on human frailties, such as arrogance and ecological malfeasance, neither of which has a nice ring but which are the real causes of our environmental problems. The term planetary management, while simultaneously appealing to our desire to be in control of the planet, avoids such messy subjects as ethos, politics, justice, and the discipline of equitable distribution of resources and of moral choice. In addition, management is a mechanical concept, one that we like because it reinforces our belief that we either are or can be in control of Nature.[8]

Ideally, management means knowing what is manageable and what is not and having the wisdom to leave the latter to manage itself. The ideal notwithstanding, our insistence on managing what cannot be sustainably managed (a tree farm or a population of salmon or tuna) while leaving unmanaged what could be managed sustainably (a forest or an ocean) is one source of the problem. Additionally, our lack of will to control our materialistic appetites, expansive economies, exploitive technologies, and exploding population is causing us to unintentionally redesign the Earth. That we must redesign the Earth is a given, simply because we exist, but how we go about redesigning the Earth is another issue altogether.

In redesigning the Earth, sustainability is a dimension in the scale of time, a dimension we ignore at our peril as we consider how we will treat our available resources. The paradigm within which our decisions are made today in economics, agriculture, and the determination of yields in forestry is based on a timeless view of reality. This economic view of quantitatively constant values, such as the notion of the unchanging biological fertility of the soil, stability of climate, and dismissal of both biological and genetic diversity as unnecessary, is both mechanistic and ecologically naïve.

Further, such concepts of sustained yield in timber, sustained yield in finances, and sustained growth in economics are all efforts to deny the existence of time and change. Evolution, in this view, is reduced to a mechanical process without any place for the novelty of change; it is a clockwork world with no capacity for a creative process. It allows only reversible locomotion devoid of creativity, like a train running endlessly back and forth on a track.[9]

The consequence of this perspective for economists is that, in their view of resources, soil, water, air, sunlight, biological diversity, genetic diversity, and climate are ignored because they are considered to be "free" and their degradation thus imposes no "costs." "Free" in this sense means they belong to no one and thus to everyone, so their quick exploitation makes sense; otherwise, their perceived short-term economic value will be lost.

So it is that the timber industry, having nothing invested in the growing of the old-growth trees, sees the old-growth forest as free and all too often ignores the degradation of the land caused by logging. The timber industry contends that the wood fiber it harvests is a product of the air and sunlight, and thus ignores the qualitative depreciation of the biological capital of the soil, which is largely governed by biological diversity, genetic diversity, and climate.

The timber industry also tends to ignore the creative evolution of the forests it harvests, evolution that is ongoing in the genetic, ecological, and cultural sense. Norman Jacob, founder of the Pacific Institute for the Study of Cultural and Ecological Sustainability, puts it this way: the "normal" forests of industrialists and the "perfect" markets of economists are timeless, qualitiless, mechanical, and strictly fictional constructs of what in reality is a timeful, qualitative, and organic world.[10]

In such a static world devoid of change and without novelty, efficiency of use in resources is an attractive concept, because it is easy to "know" what is best in a predictable, objective, static, and mechanical world. In such a world, the perspective of relationship in space and the irreversible quality of time do not exist, and experts can, therefore, "tell it like it is" instead of admitting they are "telling it as they perceive it." The economic goal thus becomes the homogenization of the creative, evolutionary process through objectivity, whereas human subjectivity, which at best is only a clear perspective from one point of view, is summarily invalidated.

Although the computer has become a metaphor for our linear understanding of such things as economics, forestry, and many other fields of endeavor, including science, we still have to contend with our human subjectivity. We still have to deal with the fact that when we optimize one set of circumstances we do not necessarily optimize another.

In a world seen as linear and locomotive, a world in which something only goes forward and backward, optimization appears to make sense and is often thought to be somehow synonymous with sustainability. In reality, however, we can neither "optimize" nor even choose what is "best" ecologically, because we live in an evolving world, where change, novelty, and uncertainty preside, where our perception of what is at any given point in time is a moving target that changes with each set of new data.

It is therefore sometimes better to do what is satisfactory, what will suffice, or even what is deemed mediocre, if that is the most we can do and feel intuitively good about it in our humanly subjective way. It is better not to *know* what is ecologically optimum or what is ecologically best in a dynamic world where the only certainty of the present is the uncertainty of the future. When we insist on optimizing or doing what we intellectually know is best ecologically, it is inevitably from an economic/political point of view, which is almost always counter-intuitive and ecologically unwise. If, however, we honor our intuition, we will find the place where culture meets Nature and both are more nearly sustainable.

Third Element: Understanding and Accepting That We Make an Ecosystem More Fragile When We Alter It

Ecosystems are designed by the *variability of the variability* of natural phenomena, such as volcanos, climate, fires, floods, and the cyclical nature of populations of organisms; they are not designed by the predictable averages of anything. Ecosystems are designed by novelty and uncertainty, not by static surety.

In addition, while incremental changes in an ecosystem may seem to us humans to be insignificant and their effects for a time to be invisible, ecosystems operate on thresholds with unknown margins of safety. But once a threshold is crossed, it is crossed. There is no going back to the original condition. It is thus necessary to understand something about the relative fragility of simplified ecosystems as opposed to the robustness of complex ones.

Fragile ecosystems can go awry in more ways and can break down more suddenly and with less warning than is likely in robust ecosystems, because fragile systems have a larger number of components with narrow tolerances than do robust ones. As such, the failure of any component can disrupt the system. Therefore, when a pristine ecosystem is altered for human benefit, it is made more fragile, which means that it will require more planning and maintenance to approach the stability of the original system. Thus, while sustainability means maintaining the critical functions performed by the primeval system, or some facsimile thereof, it does not mean restoring or maintaining the primeval condition itself.[6]

If one thus looks at ecosystems along a continuum of naturalness (the most pristine being the most natural end of the continuum and the most humanly altered being the most cultural end of the continuum), the notion of system fragility not only makes sense but also offers humanity a range of choices. And it is, after all, the array of choices that one generation passes to the next that conveys the sustainability of potential outcomes.

For example, the less we humans alter a system to meet our necessities, the more the system's functional requirements are met internally to itself. This in turn makes it easier and less expensive in both time and energy (including money) to maintain that system in a relatively steady state because we have maintained more of the diversity of native flora and fauna than we might otherwise have done.

Conversely, the more altered a system is, the more that system's functional requirements must be met through human-mediated sources external to itself. This in turn makes it more labor intensive and more expensive to keep that system in a given condition because we have maintained less, often far less, of the diversity of native flora and fauna than we would otherwise have done.

"But why," I am often asked, "do we need such a variety of species? What effect does a variety of species have on an ecosystem anyway?" One marvelous effect such a variety has is increasing the stability of ecosystems through feedback loops, which are the means by which processes reinforce themselves.

Strong, self-reinforcing feedback loops characterize many interactions in Nature and have long been thought to account for the stability of complex systems. Ecosystems with strong interactions among components, which are contributed by feedback loops, can be complex, productive, stable, and resilient under the conditions to which they are adapted. When these critical loops are disrupted, such as in the extinction of species and the loss of their biological functions, these same systems become fragile and easily affected by slight changes.

It is the variety of species that creates the feedback loops. That is what makes each individual species so valuable: each species by its very existence has a shape and therefore a structure, which in turn allows certain functions to take place, functions that interact with those of other species. All of this is governed ultimately by the genetic code, which by replicating species' character traits builds a certain amount of redundancy into each ecosystem. (For a thorough discussion of these ideas and the cause and effect of the extinction of species, see *Global Imperative.*[11])

Redundancy means that more than one species can perform similar functions. It is a type of ecological insurance policy, which strengthens the ability of the system to retain the integrity of its basic relationships. The insurance of redundancy means that the loss of a species or two is not likely to result in such severe functional disruptions of the ecosystem so as to cause its collapse because other species can make up for the functional loss. But there comes a point, a threshold, when the loss of one or two more species may in fact tip the balance and cause the system to begin an irreversible change. That change may signal a decline in quality or productivity.

Although an ecosystem may be stable and able to respond "positively" to the disturbances in its own environment to which it is adapted, this same system may be exceedingly vulnerable to the introduction of foreign disturbances (often those introduced by humans[11]) to which it is not adapted. We can avoid disrupting an ecosystem supported by feedback loops only if we understand and protect the critical interactions that bind the various parts of the ecosystem into a functional whole.

Diversity of plants and animals therefore plays a seminal role in buffering an ecosystem against disturbances from which it cannot recover. As we lose species, we lose not only their diversity of structure and function but also their genetic diversity, which sooner or later results in complex ecosystems becoming so simplified they will be unable to sustain us as a society. Therefore, any societal strategy aimed at protecting diversity and its evolution is a critically important step toward ensuring an ecosystem's ability to adapt to change. Diversity counts. We need to protect it an any cost.

Although ecosystems can tolerate cultural alterations, those functions that have been disrupted or removed in the process (often through a loss of species) must be replaced through human labor if the system is to be sustainable. The more a system is altered and simplified, the more fragile it becomes and the more labor intensive its maintenance becomes. When alterations exceed the point at which human labor can maintain the necessary functions, the system collapses.

Collapse in this case means that it becomes something other than that for which it was originally groomed, and in the process, it becomes nonproductive of that for which it was altered. The degree of human alteration determines which way a system will go, either back toward its original condition or toward something totally different.

Collapse is not always caused by dynamics internal to the system itself, however. To understand this, we need to look at one ecosystem (Easter Island[12]) that does appear to have collapsed from within as a basis of comparison with others that did not.

Self-Destruction

Easter Island is a tiny, 43-square-mile piece of land in the South Pacific 2,400 miles off the coast of South America. The oldest pollen dates on the island go back some 30,000 years, long before the first people, wandering Polynesians, arrived. At that time, the island was forested.

The Polynesians settled on the island in about the year 400 A.D. They began gradually to clear the land for agriculture, and they cut trees to build canoes. The land was relatively fertile, the sea teemed with fish, and the people flourished. Their population rose to about 15,000, and the culture grew sophisticated enough to carve the giant statues that have since become famous. The people eventually also cut trees to provide logs for transporting and erecting those hundreds of eerie statues, or *moai*, some of which are about 32 feet high and weigh as much as 85 tons.

Unfortunately, when the trees were cut, they did not grow back. Deforestation began about 1,200 years ago, a few hundred years after the people arrived, and was almost complete by 800 years ago. The people of Easter Island also exploited many of the island's other resources, such as its abundance of birds' eggs. The result was ecological disaster.

The people had cleared so much of the forest that they were without trees to build canoes for fishing. They probably also had exploited the eggs of the sooty tern to the point that the bird no longer nested on the island. And deforestation led to erosion of the soil and reduced yields of crops.

The downward spiral of culture on Easter Island had begun. Fewer fish, eggs, and crops led to a shortage of food. Hunger in turn brought warfare, even cannibalism, and the whole civilization was pushed to the brink of collapse. By the time European explorers arrived in the 1700s, only 4,000 people remained on the island, and the culture that produced the statues had completely disappeared.

Today, all we can do is marvel at the remains of the culture of Easter Island—statues that once stood erect on specially built platforms, others that lie abandoned between the volcanic quarries of their origin and their planned destinations, and still others that remain unfinished in the quarries.

Easter Island is an example of habitat-dependent culture that caused the collapse of its ecosystem from within. Now let's see what can happen from without.

Loss of Labor Pool

In contrast, let's look at ancient Greece.[6,13] Greece, flourishing under wise agricultural use during the beginning of the Iron Age, had nevertheless greatly altered its landscape, despite its apparently sound agricultural ethic. But all the human-caused changes, including deforestation, do not appear to have caused the collapse of the agricultural system. In fact, it was not only sustainable and was being sustained but also might have continued to the present day if it had not been for the effect of outside influences.

While the Greeks modified their landscape, making it fragile, their agricultural system was sustainable as long as there was a full human population to tend to the terraced fields. The destruction of their agricultural system was not a consequence of the system itself, but rather of Romans raiding the Greek countryside for slaves, which reduced the population and left the fragile landscape untended to wash into the sea.

Thus, as long as the Greeks maintained adequate cover crops, which were both labor intensive and functioned to hold in place the soil as the forests had once done, their agricultural system was sustainable. But as the activities of Roman slavers continually reduced the Greek population, there came a threshold beyond which this labor-intensive agriculture simply could not be maintained, and the system collapsed.

Prior to the advent of Greek agriculture, the land had been forested for millennia, making sustainability a moot point. Sustainability arose as a problem not because of deforestation, but because of the inability of a society debilitated by slaving to continue performing the function of the forest, namely soil conservation.

This same kind of dynamic is occurring today in Malaysia and many other parts of the world, but for another reason. While working in peninsular Malaysia, I observed a number of abandoned rice paddies, some of which were being reclaimed by young-growth jungle while others were simply eroding away. When I asked why this was happening, I was told that many of the younger people were migrating to the cities, such as Kuala Lumpur.

Growing rice without modern machinery is labor intensive. As long as there are enough young people in the villages to augment and eventually replace the old people in the labor pool, the rice paddies will be sustainable. But as the young people leave the villages for the cities, they diminish the village labor pool just as surely as the Romans did when they captured and removed Greek peasants as slaves. When a village labor pool falls below a certain threshold minimum, the rice paddies are no longer sustainable. Now consider the introduction of technology.

Introduced Technology

In addition to a physical loss of individuals from a communal labor pool, introduced technology, which replaces traditional human labor without removing the people from the potential labor pool, can also cause an internally sustainable social/environmental system to collapse. In June 1992, I visited Čergov, in northern Slovakia, to evaluate the condition of the native forest, which is primarily European beech with an admixture of white fir. The native forest was being rapidly clearcut and replaced with plantations of such nonnative species as Norway spruce, larch, and pine. The biological errors of forestry made in Germany, the United States, and Canada (replacing biologically diverse forests with short-lived simplistic economic tree farms) were being repeated in the forests of Čergov and for the same reasons— immediate and short-term economic gain.

Among the most graphic examples of technology disrupting a centuries-old sustainable system of selective logging was the recent clearcutting and its resultant loss of topsoil. Within one hour after each thunderstorm in Čergov, all the streams and rivers fed by clearcut slopes went from clear water to the appearance of milk chocolate as the soil of the forest was washed away to the sea.

Prior to importing the technology of clearcutting, the forest of Čergov was selectively logged with horses. In addition, horse logging had been biologically sustainable for centuries, as were the economies of the small mountain villages located in the upper valleys near the edge of the forest. Now the jobs once sustained by horse logging are gone, like the topsoil of the forest. The people who once made their living from the forest must commute to the cities to find work, and the villages have lost part of their cultural heritage.

Once outside desires, such as economics, become intertwined in the balance of an internally sustainable social/environmental system, it is easy to ignore and thus lose sight of a sound ecological perspective. For example, I was told by a forester from a distant city that it was inevitable that the old trees of Čergov would be clearcut and that plantations of spruce would replace them. I questioned this statement because there was nothing inevitable about either clearcutting the old trees or planting spruce in rows. It was simply someone's choice to maximize immediate economic gain.

The forester told me that clearcutting was necessary because the ground was too steep for logging with horses. I found this statement to be particularly interesting because the forest had been logged with horses for centuries before the chain saw and log truck became available, and all the clearcuts I saw could easily have been logged with horses. Although such technology as chain saws and log trucks has the ability to directly disrupt local ecosystems, the effects of technology can grossly alter ecosystems at great distances from the technological apparatus itself.

Long-Distance Transport of Air Pollutants

We pollute the air with chemicals. Air pollution directly affects vegetation by altering the quality of the soil and water as well as the quality and quantity of the sunlight that drives the plant/soil processes. The chemicals we dump into the air also alter the climate and thus the environment in which the vegetation grows.

Consider acid rain, for example. Acid rain has long been recognized as a pollution problem in Europe, where statues and gargoyles that once proudly adorned city streets and plazas and guarded centenarian buildings have had their faces dissolved over recent decades. The statues that I remember seeing as a boy, in perfect form and feature, today are often-unrecognizable relics of a past era because acid rain has eaten away the marble much as leprosy eats away the flesh.

Acid rain is not confined to European cities, however. It is also found in forest and fen, in highland and lowland. There, too, it is destroying the essence of life as it joins league with other forms of industrial/technological pollution, where it contributes to a phenomenon the Germans call *Waldsterben*, the dying forest.

The dying forest syndrome is not exclusively the property of Europe; it is owned by every industrial country, including the United States and Canada. Here, called forest dieback, it manifests itself primarily along the eastern seaboard, where declining growth rates and the dieback of red spruce and other species of trees, particularly at high elevations, are attributed to atmospheric pollution, of which acid deposition is one of the most widespread components.

A primary human source of the precursors to acid deposition is coal-fired power plants, which still provide more than 55 percent of the electricity generated in the United States and account for about one third of the nitrogen oxides and about two thirds of the sulfur dioxide produced each year.[14] These atmospheric pollutants are capable of a phenomenon known as long-distance transport, which simply means that they can travel great distances from their sources on air currents (even as far away as the arctic and Antarctica) before being deposited on agricultural fields and forests, where they affect plant growth.

Direct effects of acid deposition include a reduced functioning of plant roots. This reduction in root function may be caused by chemical changes in the soil as a result of acid deposition, by reduced translocation of carbohydrates from pollution-damaged shoots, or from excessive nitrogen.[15,16] In fact, the emission of pollutants has tripled the level of nitrogen some forests are receiving.[17]

Change in nitrogen level works as follows: People are used to the idea that if one puts nitrogen on a corn field, the corn grows faster, but then corn happens to be a plant that opens its stomata (pore-like entrances into the leaves) and takes up more carbon to create a balance. Thus, up to some point, the more nitrogen corn is given, the more carbon it will acquire, and the better it will grow. Unlike corn, however, most forest species will not grow any faster with additional nitrogen.[17]

Since trees do not take up any more carbon to offset the additional nitrogen, an altered carbon–nitrogen ratio develops in the plant tissue, which means that trees are receiving three times the level of their nitrogen tolerance. This *does not* mean that they are getting three times their optimum level, but rather three times their *tolerance.* Such a shifted carbon–nitrogen ratio translates into, among other things, an alteration of the materials plants produce to resist diseases and insects, which permits pathogenic fungi (those that create rot in living trees) to enter where previously they had been excluded.[17]

How might this undermine cultural sustainability? The people of a Canadian First Nation (which we would think of as a tribe of Indians) are living sustainably within their ancestral lands, which lie within the northern edge of the boreal forest. Their dominant means of livelihood is hunting, fishing, and trapping fur-bearing mammals, as they have done for centuries. Being isolated by miles of unbroken forest and being closely tied to their ancestral lands, they have largely stayed within their borders.

Enter airborne sulfur dioxide from the cities of southern Canada and the northern United States, well over a thousand miles away, and the resulting acid deposition begins in secret to kill the boreal forest. The demise of the forest is at first invisible, but then the people begin to notice that something is wrong. Gradually, the condition of the forest deteriorates to the point that it alters the habitats of the animals on which the people have traditionally relied for food and as money for trade.

With the decline in forest health comes poverty and destabilization of the people's culture within their ancestral home. With nowhere else to go, the culture collapses with the collapse of the forest's biological sustainability, through no fault of the First Nation peoples. In fact, they have not a clue as to what happened. All they know is that the forest, whose sustainability they have respected, protected, and relied on for centuries, is suddenly inexplicably dying and they are dying with it.

In addition to acid rain, an analysis of tree bark, collected from the tropics to the chilled latitudes, indicates that insecticides and fungicides have spread and are spreading around the world on air currents. Traces of chemicals found in the bark include those related to DDT, lindane, chlordane, aldrin, and more than a dozen other insecticides and fungicides. Some of these chemicals become airborne in hot climates and are carried to cooler areas, where they condense out of the atmosphere and concentrate far from where they were sprayed, often thousands of miles away. Some were even sprayed decades ago but are still affecting the environment[18] through the pollution of soil and water.

Direct and Indirect Pollution of Soil and Water

> Phosphate fertilizer factories emit toxic fumes and build mountains of mildly radioactive waste as earth-moving machines scar the earth.
>
> Other industries sink narrow pipes like giant hypodermic needles and inject millions of gallons of poisonous liquids deep into the earth, beneath Florida's main supply of drinking water.
>
> Paper mills dump toxic wastes into rivers and bays, contaminating fish, spoiling wells.
>
> This is the other Florida, away from the sunny beaches, horse tracks, Walt Disney World, resort hotels and designer golf courses that draw millions of tourists a year.[19]

Unintentional fragility is also imposed on ecosystems through both the direct and indirect pollution of soil and water. Soil, which is like an exchange membrane between the living (plant and animal) and nonliving components of the landscape, is dynamic and ever-changing. Derived from the mechanical and chemical breakdown of rock and organic matter, soil is built up by plants that live and die in it. It is also enriched by animals that feed on plants, void their bodily wastes, and eventually die, decay, and return themselves to the soil as organic matter. Soil, the properties of which vary from place to place within landscapes, is by far the most alive and biologically diverse part of a terrestrial ecosystem. In addition, soil organisms are the regulators of most processes that translate into soil productivity.

The soil food web is a prime indicator of the health of a terrestrial ecosystem. But soil processes can be disrupted by such things as decreasing bacterial or fungal activity, decreasing the biomass of bacteria or fungi, altering the ratio of fungal to bacterial biomass in a way that is inappropriate to the desired system, reducing the number and diversity of protozoa, and reducing the number of nematodes and/or altering their community structure. Such disruptions can lead to a loss of vegetation or even the loss of human health.[20]

Soil, which is the main terrestrial vessel, receives, collects, and passes to the water all airborne, human-caused pollutants. In addition, such pollutants as chemical fertilizers, fungicides, herbicides, insecticides, rodenticides, and so on are added directly to the soil and through the soil to the water. At times, such pollutants make their way into the air and hence are redistributed more widely over the planet's surface through strong winds, which carry aloft the topsoil following deforestation, desertification, and ecologically unsound farming practices, and ultimately affect water.

Most of our usable water, which is a captive of gravity, comes from snows high on forested mountain slopes. When snow melts, the water percolates through the soil. It is purified when flowing through healthy soil; it is poisoned when flowing through soil stripped of Nature's processes and polluted with artificial chemicals. In addition, water bearing tons of toxic effluents flows directly into streams, rivers, estuaries, and oceans.

Water, the great collector of human-caused pollutants, washes and scrubs the pollutants from the air by rain and snow; it leaches them from the soil, and it carries them in trickle, stream, and river to be concentrated in the ultimate vessel, the combined oceans of the world.

Humans directly affect air and both directly and indirectly affect soil and water. If, for example, we choose to clean the world's air, we will automatically clean the soil and water to some extent because airborne pollutants will no longer poison them. If we then choose to treat the soil in such a way that we can grow what we desire without the use of artificial chemicals and if we stop using the soil as a dumping ground for toxic wastes and avoid overintensive agriculture, the soil can once again purify water by filtering it. If we then stop dumping waste effluents into ditches, streams, rivers, estuaries, and oceans, they can again become clean and healthy.[21]

With clean and healthy air, soil, and water, we can also have clear, safe sunlight with which to power the Earth and, with the eventual repair of the ozone shield, a more benign—and perhaps predictable—climate in which to live. In addition, effective population control can tailor human society to fit within the world's carrying capacity.

A population in balance with its habitat will reduce demands on the Earth's resources. With reduced competition for resources can come the cooperation and coordination that will allow our landscapes to provide the maximum possible biodiversity. Protecting biodiversity translates into the gift of choice, which in turn translates into hope and dignity for future generations.

If we do everything outlined here except clean the air, we will still pollute the entire Earth. *Clean air is the absolute bottom line for global biological sustainability and therefore human survival.* Without clean air, there eventually will be no difference in the way we destroy ourselves, either by genocide or air pollution (indirect suicide), because our biosphere is comprised of interrelated, interdependent, interactive components, where one affects the whole and the whole affects the one. But there are alternatives, based on intrinsic values. The choice is ours. To the children we bequeath the consequences.

Fourth Element: Understanding and Accepting That We Must Reinvest in Living Systems Even as We Reinvest in Businesses

The industrialized nations of the world operate with the idea that for an economic endeavor to be healthy it must forever expand. Thus we attack the world's renewable natural resources from an exploitive point of view, which increasingly poses limits on most, if not all, "renewable" resources.

The timber industry, a prime example, operates in a perpetual expansionistic mode, particularly in areas where considerable native forest remains. As a result, the world's often irreplaceable forested resources are rapidly shrinking. Perpetual expansion involves liquidating the native forest and ultimately exhausting the soil.

The economic concept of waste, which excludes nonmonetary social values and all intrinsic ecological values, has spawned the industrial concept of salvage logging, usually in the form of clearcutting. Clearcutting, in turn, is an economic expediency in which I find no biological justification—it mimics nothing in Nature.

According to the 1981 *American Heritage Dictionary*, salvage means to save any imperiled property from loss or destruction. Salvage logging, which views trees only as imperiled economic property, exemplifies the philosophy of traditional forestry, which is the belief that any potentially merchantable tree left in the forest to rot is an economic loss, a waste. But in a biological sense, there is no such thing as waste in a forest.

A tree rotting in the forest is a *re*investment of Nature's biological capital in the long-term maintenance of soil productivity, and hence the forest itself. In a business sense, one makes money (economic capital) and then takes a percentage of the earnings and reinvests them in the maintenance of buildings and equipment so as to continue making a profit by protecting the initial investment over time. In a business, one reinvests after the fact, after the profits have been earned.

It is different, however, with biological capital, which is the capital of all renewable natural resources. A forest, for example, simply cannot process economic capital; it requires such biological capital as organic material and biological and genetic diversity. In a forest, one must reinvest before the fact by leaving some proportion of the merchantable trees in the forest to rot and thereby reinvest themselves in the fabric of the living system.

Such biological reinvestment, including large merchantable trees, both live and dead, is *necessary* to maintain soil health, which in large measure equates with the health of the forest. The health of the forest, in turn, equates with the long-term economic health of the timber industry and therefore of some human communities.

Planting and fertilizing trees are not *re*investments, as commonly held by the timber industry, but rather investments in the next commercial crop of trees. As such, they are investments in a potential product, not in the biological sustainability of the living system that produces the product. We do not reinvest in maintaining the health of biological processes because we do not see the forest; we only see the commercial product. We do not reinvest because we insist that ecological variables are really constant values that we need not consider and can even discount.

"What," you may ask, "will we lose if we convert forests into economic plantations (tree farms) through such practices as clearcutting and salvage logging?" Before answering this question, it is important to understand that all things in Nature's forest are neutral when it comes to any kind of human valuation. Nature has only intrinsic value. Thus, each component of the forest, whether a microscopic bacterium or a towering 800-year-old tree, is therefore allowed to develop its prescribed structure, carry out its prescribed function, and interact with other components of the forest through their prescribed interrelated, interactive, interdependent processes. No component is more or less valuable than any other; each may differ in form, but all are complementary in function.

A native old-growth forest, for example, has three prominent characteristics: large live trees, large standing dead trees or snags, and large fallen trees. The large snags and the large fallen trees become part of the forest floor and eventually are incorporated into the forest soil, where myriad organisms and processes make the elements stored in the decomposing wood available as nutrients to the living trees. These processes are all part of Nature's rollover accounting system, which includes such assets as large dead trees, genetic diversity, and biological diversity, all of which count as reinvestments of biological capital in the growing forest.

Intensive short-term plantation management (tree farms) disallows reinvestment of biological capital in the soil and therefore in the forests of the future, because such reinvestment has come to be seen, erroneously, as economic waste. We therefore plan the total exploitation of any part of the ecosystem for which we see a human use, and we plan the elimination of any part of the ecosystem for which we cannot see such a use. With this myopic view, we eliminate the diversity of Nature through our existing economic planning system, which inevitably leads to biological extinction of species and their functions within the ecosystem (witness the rapidly growing number of endangered species in the world).

After a native forest is liquidated, we may be deceived by the apparently successful growth of a first-rotation tree farm, which lives off the stored, available nutrients and processes embodied in the soil of the harvested native forest. Without balancing biological withdrawals, investments, and reinvestments, both biological interest and principal are spent, and so both biological and economic productivity must eventually decline.[22,23]

Consider, for example, the coniferous forests of the Pacific Northwest in which Douglas fir and western hemlock predominate in the old-growth canopy. Herein lives the spotted owl, which preys on the flying squirrel as its stable diet. The flying squirrel, in turn, depends on truffles, the belowground fruiting bodies of mycorrhizal fungi. (The term mycorrhiza, which means fungus root, denotes the obligatory symbiotic relationship between certain fungi and plant roots.) Flying squirrels, having eaten truffles, defecate live fungal spores onto the forest floor, which, upon being washed into the soil by rain, inoculate the roots of the forest trees. These fungi depend for survival on the live trees, whose roots they inoculate, to feed them sugars, which the trees produce in their green crowns. In turn, the fungi form extensions of the trees' root systems by collecting minerals, other nutrients, and water, which are vital to the survival of the trees. Mycorrhizal fungi also depend on large rotting trees, lying on and buried in the forest floor, for water and the formation of humus in the soil. Further, nitrogen-fixing bacteria occur inside the mycorrhiza, where they convert atmospheric nitrogen into a form that is usable by both the fungus and the tree.

Such mycorrhizal/small mammal/tree relationships have been documented throughout the coniferous forests of the United States (including Alaska) and Canada. They are also known from Argentina, Europe, and Australia.

All this is complicated, but so is Nature's forest. To add to the overall complexity, a live old-growth tree eventually becomes injured and/or sickened and begins to die. How a tree dies determines how it decomposes and reinvests its biological capital (organic material, chemical elements, and functional processes) back into the soil and eventually into another forest.

A tree may die standing as a snag, to crumble and fall piecemeal to the floor of the forest over decades, or it may fall directly to the floor of the forest as a whole tree. Regardless of how it dies, the snag and fallen tree are only altered states of the live tree; the live old-growth tree must therefore exist before there can be a large snag or fallen tree.

How a tree dies is important to the health of the forest because its manner of death determines the structural dynamics of its body as habitat. Structural dynamics, in turn, determine the biological/chemical diversity hidden within the tree's decomposing body as ecological processes incorporate the old tree into the soil from which the next forest must grow.

What goes on inside the decomposing body of a dying or dead tree is the hidden biological and functional diversity that is totally ignored by economic valuation. That trees become injured and diseased and die is therefore critical to the long-term structural and functional health of the forest, but to an industrial forester such injured and diseased trees are seen only as an economic waste if not cut and converted into money.

The forest is an interconnected, interactive, organic whole defined not by the pieces of its body but rather by the interdependent functional relationships of those pieces in creating the whole—the intrinsic value of each piece and its complementary function. These functional relationships are totally ignored in salvage logging.

Let's return for a moment to the Pacific Northwest, where the spotted owl preys on the flying squirrel, which depends on truffles for its diet. The fungus, of which the truffle is a part, is closely associated with large wood on and in the forest floor. The squirrel, the owl, and the fungus all depend on the same wood!

Salvage logging disrupts this interconnected, interactive, interdependent relationship by removing all merchantable dying and dead trees. The danger of such logging lies primarily in its philosophical underpinnings that justify immediate economic considerations to the exclusion of all else. Members of Congress in favor of such economic tactics recently attached authority for salvage logging as a rider to an unrelated bill. The purpose of this maneuver was to insulate short-term monetary profit from environmental law and any legal challenge to lawbreakers.

Salvage logging, as currently practiced, has the following immediate consequences:

1. Areas where logging has heretofore been prohibited (i.e., roadless areas) will be opened to roads, thus destroying forever their integrity as roadless areas.
2. Arson fires will probably increase to stimulate salvage sales as a means of logging as much of the remaining old-growth forest as possible.
3. Timber will most likely be salvaged through clearcutting, which is a drastic biological simplification of a complex forest ecosystem.
4. Salvage, as normally practiced, mimics clearcutting, and clearcutting mimics nothing in Nature.
5. Normal logging is designed to make money, but within at least some planned ecological constraints. Salvage logging is reactive to keep from losing possible monetary gains and is thus unplanned, opportunistic, and without ecological constraints.

6. Normal logging compacts soil and removes a pre-established volume of timber, theoretically within some ecological constraints. Salvage logging is a re-entry of logged sites, which further compacts the soil, nullifying any ecological constraints.
7. Initial logging is most often based on what is and is not to be cut. Salvage logging opens the real possibility of individual on-site interpretation of what to cut, including such things as live "risk trees" or live "associated trees."

Cutting some dying and dead trees for commercial use is neither morally wrong nor necessarily ecologically harmful. But the philosophy and practice of salvage logging is the epitome of nonsustainability because it employs overriding short-term economic rationale as an excuse to summarily ignore all current ecological knowledge about the long-term biological sustainability of forests. The sole objective of salvage logging is to convert trees into money, thus replacing the art of forestry with the technology and economics of cutting trees.

Traditional forestry practice, now outmoded because we have improved information, began with the idea that forests (considered only as collections of trees) were perpetual economic producers of wood. With such thinking, it was necessary to convert a tree into some kind of potential economic commodity before it could be assigned a value.

The rationale for converting trees into money came from the "soil-rent theory." "Soil rent," devised in the early 19th century, fit perfectly with classic liberal economic theory, which was designed to maximize profits as the general objective of economic activities. Since its adoption by foresters, it has been the general overriding objective of industrial forestry as well.[23]

This economic theory, however, is based on six greatly flawed assumptions: (1) the depth and fertility of the soil in which the forest grows is nondegradable, (2) the quality and quantity of precipitation reaching the forest is unchanging, (3) unpolluted air infuses the forest, (4) diversity (biological, genetic, and functional) is unimportant, (5) the amount and quality of solar energy available to the forest are constants, and (6) climate is unchanging.

Erroneously assuming that ecological variables can be considered economic constants leads to the further false assumption that Nature recognizes the economic notion of an "independent variable." The concept of an independent variable means that soil, water, air, biodiversity, etc. can be considered constant while manipulating a single desired economic entity—the tree. If there were in fact such a thing as an independent variable, biological sustainability for any tree species could be calculated using only two considerations: rate of growth and age at which the tree must be cut in order to gain the highest rate of economic return in the shortest time for the least investment.

The potential for converting trees and other resources into money counts so heavily because the economically effective horizon in most economic planning is only about five years away. Thus, in traditional linear economic thinking, any merchantable tree that falls to the ground and reinvests its biological capital into the soil is considered an economic waste (i.e., it has not been converted into money).

Forests are being decimated the world over because "conversion potential" dignifies, with a name, the erroneous notion that unharvested resources have no intrinsic value and must be converted into money before any value can be assigned. These notions predominate in the "management" of forests: anything without monetary value has no value, and anything with immediate monetary value is wasted if left unused by humans.

We cannot, however, have an economically sustainable yield of any forest product, such as wood fiber, water, soil fertility, or biological or genetic diversity, until we first have a biologically sustainable forest, one in which the biological divestments, investments, and reinvestments are balanced in such a way that the forest is self-maintaining in the long term.

Sustainability is thus additive. We must have a biologically sustainable forest to have a biologically sustainable yield, a biologically sustainable yield to have an economically sustainable industry, an economically sustainable industry to have a sustainable economy, a sustainable economy to have culturally sustainable communities, and culturally sustainable communities to have a sustainable society.

When sustainability is put in purely economic terms, the additive economic relationship of the biological yield becomes clear: We must first practice sound "bio-economics" (the economics of maintaining a healthy, sustainable forest), before we can practice sound "industrio-economics" (the economics of maintaining a healthy, sustainable forest industry), before we can practice sound "socio-economics" (the economics of maintaining a healthy, sustainable community and society).

It all begins with a solid foundation—in this case, a healthy, biologically sustainable forest.

Traditional forest practices are counter to sustainable forestry, because instead of training foresters to care for forests, we train plantation managers to manage the short-rotation *economic* plantations with which we are replacing our native forests. Forests have evolved through the cumulative addition of structural diversity, which initiates and maintains the diversity, complexity, and stability of ecological processes through time. We are reversing this rich building process by replacing native forests with plantations designed only with narrow, short-term, economic considerations.

Every acre on which a native forest is replaced with a plantation is an acre that is purposely stripped of its biological diversity, its biological sustainability, and is reduced to the lowest common denominator—simplistic economics.

Thus, the concept of a plantation or tree farm, a strictly economic concept, has nothing whatsoever to do with the biological sustainability of forests. Under this concept, native forests are replaced with plantations of genetically manipulated trees accompanied by the corporate/political promise that such plantations are not only sustainable but also better, healthier, more viable, and more productive of wood fiber than are the native forests, which evolved with the land over millennia.

But "sustainable" means producing industrio-economic outputs while maintaining the viability of forest processes (bio-economics) in perpetuity. In turn, this necessitates balancing withdrawals of products with bio-economic reinvestments in the health of the forest, especially the soil. It means maximizing the health of the forest and using all products and enjoying all amenities thereof with humility and gratitude.

Fifth Element: Understanding and Accepting That Only a Unified Systemic World View Is a Sustainable World View

The antecedent to and foundation of our world view, which is a society's current operating paradigm of how the world works, was fashioned by such rationalist thinkers as Francis Bacon (1561–1626, English philosopher and essayist), Galileo Galilie (1564–1642, Italian scientist and philosopher), René Descartes (1596–1650, French philosopher and mathematician), John Locke (1632–1704, English philosopher), Isaac Newton (1642–1727, English mathematician, scientist, and philosopher), and Adam Smith (1723–1790, Scottish political economist and philosopher).

Consider the collective paradigm of these renowned men: Nature's sole value is in service to the material desires of humanity (Bacon). But Nature must be tortured before Her secrets will be revealed for human use (Bacon). Once wrested from Nature, only those secrets that are measurable and quantifiable are real or relevant and can be studied (Galilie). Because real things are both measurable and quantifiable, they must operate through predictable linear mechanical principles, like an enormous machine (Newton). And like a machine, real things can be understood by disassembling the things themselves into smaller and smaller, more manageable pieces, which can then be rearranged in an order deemed logical to the human mind (Descartes).

With reductionistic mechanical logic, major segments of Western society confer upon themselves the unlimited rights of individual private property (a prostitution of Locke's original intent) for which people must compete with one another in pursuit of their own self-interests (Smith). (*Reductionistic* is used in the sense of taking apart and isolating the components of a living system, rearranging them according to human logic, keeping only those deemed of value in human terms, reassembling the retained pieces, and expecting the system to function as before. This is much like dissecting a cat, keeping the parts for which one sees a value, arranging the parts in a humanly conceived order, sewing them together, and expecting the cat to live and behave like it did.[24] *Mechanical* refers to the common human notion that the world is assembled like a machine, acts like a machine, and thus can be treated like a machine, which has interchangeable parts, like a watch.)

Such self-interest is to be free from any government interference because the "Invisible Hand" of moral guidance will temper self-interest in the pursuit of material wealth—for the betterment of society (Smith). While Smith's "Invisible Hand" may have had spiritual connotations, they are ignored in the current pursuit of self-interests in the form of material wealth. Further, his notion of a Higher Moral guiding human action was already overshadowed by the accepted reductionistic mechanical posits of Bacon, Galilie, Descartes, and Newton.

Consider, for example, rancher Jones, who has a stream running through one edge of his property. The property of rancher Thomas, Jones's neighbor, abuts Jones's property about 50 feet from the stream.

Downstream of Jones's property, the stream runs through Harry's Trout Farm.

Thomas's property near the stream is a large wet meadow, which is very important as a pasture to his cattle because the meadow is irrigated belowground by a high water table maintained by the stream's lowest summer level of water. And Harry's Trout Farm relies on the clean, cold water of the stream both for healthy fish for customers who want to angle and for the fish eggs Harry collects and rears each year to replenish his stock.

Jones, an old man, dies suddenly and his ranch is put up for sale. Mr. Gobel, a totally self-centered man, purchases it, and his first two acts are to get rid of the fences protecting the stream's banks and to double the number of cattle on the ranch. With the fences gone, the cattle spend most of their time along and in the stream, where they immediately begin destroying the vegetation that shades and cools the water and breaking down the banks, which starts a self-reinforcing feedback loop of erosion. While erosion commences adding silt and organic debris to the stream's once clear water, the loss of the stream-side vegetation increases the water's temperature.

Within a year, both Thomas and Harry visit Mr. Gobel and ask him to please protect the stream again, as rancher Jones had done. Without protection of the stream, they tell him, the water table is beginning to sink as the stream's channel erodes downward, resulting in the drying of the meadow. This will destroy it as pasture for Thomas's cattle, something he cannot afford. And Harry's trout business is declining because there is now so much organic debris in the water that the eggs he harvests succumb for lack of oxygen, and the adult fish cannot tolerate the increasing water temperature in summer. This is going to put him out of business.

Mr. Gobel's response to his neighbors' entreaty on behalf of the stream's health is to say: "This is my property. I have clear title to it, and by God I'll do with it as I see fit. Who do you think you are trying to tell me what to do with my own place. Get off *my* land and stay off!" As a consequence of Mr. Gobel's *rights of private property*, both Thomas and Harry are put out of business, and Mr. Gobel purchases their land.

In the end, however, Mr. Gobel destroys his own ranch and is himself put out of business. Now the land is so poor that no one can make a living off of it. Mr. Gobel has, in effect, used his *rights of private property* to steal from the future also.

If this example sounds farfetched, it isn't. People like Mr. Gobel act out of fear of not having enough to feel secure and out of ignorance of the consequences of their actions, not out of intentional malice. They have no concept of a partnership with Nature or their neighbors. As far as they are concerned, they have a "God-given right" to dominate and subjugate Nature for personal gain. (For further discussion of these ideas, see *Resolving Environmental Conflict.*[25])

Mr. Gobel's view of his "God-given right" to dominate and subjugate Nature for personal gain on his own property is part and parcel of the mind/body split and the human/Nature separation. Unbeknownst to Mr. Gobel, however, his view is the legacy of the reductionistic mechanical foundation of our expansionist economic world view, a view that is ever increasing our sense of isolation from one another and from Nature. This dualism has led us to treat Nature as a commodity from which we are independent and separate. By separating ourselves from Nature, we have justified our trying to control the uncontrollable.

Our analytical perspective involves a four-part process: (1) dissect the system into its component parts, (2) study each part in isolation, (3) glean a knowledge of the whole by studying its parts, and (4) rearrange the parts in such a way that they satisfy our human sense of logic based on ever-expanding capital gain.

The implicit assumption is that systems are aggregates of interchangeable parts that function in a linear fashion. Thus, by optimizing each part, we optimize the whole and continually fragment our problems into smaller, more "manageable" pieces, while our challenges are increasingly systemic. (*Linear,* in the sense it is used here, means having only one dimension, that of an ever-extending straight line with no means of return, such as an economy that is ever expanding.)

Consider, for example, a forest. A forest is dissected into its components as best we are able, such as trees, shrubs, herbs, birds, mammals, water, soil, and so on. Each component is then studied in relative isolation of the others, often in terms of its perceived economic potential. With considerable knowledge of those parts we favor for their potential economic value and almost no knowledge of those parts for which we find no immediate use, we cobble together our sense of the whole as a functioning system. Finally, we select and arrange the pieces for which we perceive an economic value in the order that we think will accrue the greatest financial benefits, in the shortest possible time, at the least possible cost, into an unlimited future.

And behold! We have converted a biologically sustainable, complex forest into a biologically nonsustainable, economically simplistic tree farm.

We have thus reduced the forest to our conception of its individual components, which we "divide to conquer" (*reductionistic thinking*). Next we select favored pieces and reassemble the system, often substituting parts, such as an exotic species of tree for indigenous ones or genetically engineered trees for wild ones, as though they were interchangeable, without any conceivable biological consequences in the present or future (*mechanistic thinking*). Finally, we plan for a sustain*ed* yield of crop trees from the tree farm into an unlimited future because the only variable we are willing to recognize is the tree we plant, which in economic theory is an independent variable. The fact that all other ecological variables have been converted on paper into economically constant values to artificially create the independent variable is discounted (*linear thinking*).

Today, fragmentation based on linear reductionistic mechanical thinking, which looks at parts and ignores the whole, continues to disintegrate our social structure by obliterating the sense of a society as a living system. This kind of fragmentation led quantum physicist David Bohm to say: "Starting with the agricultural revolution, and continuing through the industrial revolution, increasing fragmentation in the social order has produced a progressive fragmentation in our thought." Fragmentation—specialization, special interest groups, and political lobbyists—is the very underpinning of today's professionalism, and yet it is making our society increasingly ungovernable. The triumph of such reductionist thinking has given rise to a whole set of economic conditions under which we try to operate in isolation from the system itself.

Although sustainability is only possible with the acceptance of a unified world view, management, which is a reductionistic mechanical concept, is the antecedent of an expansionist economic world view. Plans to manage the Earth are therefore founded on the belief that ignorance is a problem that can be cured with science and technology, both of which are linear, reductionistic, and mechanical. And the purpose of so-called land management is to benefit an ever-expanding economy.

A world view has incredible power in directing the way people think. It often seems, for example, that the assumptions embodied in our world view have become so much "second nature" that we are unaware they exist. We perceive them to be absolute truths when in reality they are illusions accepted as truths. We must therefore examine carefully our present, obsolete, expansionist economic world view and begin a conscious transition to a world view with a far greater sense of social/environmental sustainability (a unified systemic world view) if our society is to survive the twenty-first century in any semblance of its current order.

Our behavior, the extension of our feelings, thoughts, and values, is out of harmony with both society and the environment because we insist on applying the reductionistic mechanical foundation of our current world view as it was founded over 300 years ago. Although this foundation, which once was a world view in itself, seemed correct in its time and place (as to many people does our current expansionist economic world view), the more we study the problems of our time, the more apparent it becomes that our social/environmental imbalance is a spiritual crisis. This crisis is brought about by our clinging to the expansionist economic world view and its associated value system, which has spawned profoundly lopsided and unhealthy ideologies, technologies, institutions, and lifestyles.

Whatever we discuss, be it disease, child abuse, crime, economics, energy shortages, extinction of species, forestry, pollution, or nuclear power, the dynamics are the same: an underlying crisis of perception. The values and assumptions of the expansionist economic world view on which we base our decisions are irrelevant to both the present and the future. Yet our continued acceptance of this world view as the absolute truth and the only valid way to acceptable knowledge has led to our current global crises and is propelling us ever closer to environmental destruction.

It should come as no surprise, therefore, that our attempts to "overcome these problems" are part of the exact mindset that created them in the first place. Put differently, yesterday's solutions to yesterday's problems have become today's problems, and today's solutions to today's problems, such as fragmentation, competition, and reactiveness, will become tomorrow's problems. Today's problems are not solvable with our obsolete world view because it is based on dysfunctional thought patterns concerning our current issues and must be transcended.

Today, for instance, we are so focused on our national and personal security that we do not see the price we pay for living in and with dysfunctional, self-serving bureaucratic organizations. We not only are losing the open, emotionally safe places to interact with the ever-changing patterns of life but also are continuing to ask the same old tired questions and thereby limit our imagination and new possibilities.

The Expansionist Economic World View

Professor Duncan Taylor has offered a thoughtful essay on the origins of our expansionist economic world view.[26] Much of the following discussion is based on his essay.

With a foundation of Bacon's inductive method to control Nature for human gain and the Cartesian–Newtonian world view of a great machine subject to methodologies predicated on reductionism, quantification, and the separation of facts and values came both the Enlightenment and the Industrial Revolution.

The Enlightenment, a philosophical movement of the eighteenth century, was concerned with the critical examination of previously accepted doctrines and institutions from the point of view of rationalism. The Enlightenment's faith was in the wedding of science and technology as a means of ending human scarcity and suffering. Then, in the middle of eighteenth-century England, came the Industrial Revolution with its tremendous social and economic changes brought about by the extensive mechanization of production systems, which resulted in a shift from home manufacturing to large-scale factory production. The Industrial Revolution's technological advances helped equate "progress" with the satisfaction of material wants, which eventually created a consumer-oriented society.

The Enlightenment and the Industrial Revolution, which grew out of the reductionistic mechanical world view of the time, cleared the way for a new world view—the expansionist economic world view. The expansionist economic world view, parented by capitalism, legitimized and institutionalized the lust for material wealth over which feudal society had for so long fought.

Then, associated with the West's historical experience of geographical expansion, came the concept of continuous growth based on what seemed to be boundless opportunities in a world of limitless natural resources—the personification of the expansionist economic world view. The expansionist economic world view, once entrenched, was justified as social development, which came to be thought of as social progress.

In the expansionist economic world view, Nature is seen as a vast storehouse of natural resources whose sole value lies in their conversion to commodities for the satisfaction of the ever-increasing material wants of an ever-growing human population. Consequently, material growth is equated with economic growth, which is equated with social development, which is thus seen as a prerequisite for human happiness and prosperity, the pursuit of which is built into the very fabric of American society and considered inviolate. And because material/economic growth is seen as a necessary prerequisite for human happiness and prosperity, a drop in the rate of growth is thought to inevitably result in economic stagnation, mass unemployment, and therefore social distress.

Our social predicament is a legacy of our expansionist economic world view, which finds value only in those material things that can be measured and quantified and discounts all spiritual things that defy material valuation. We therefore live in frantic pursuit of fulfillment, but a fulfillment based on substitute gratifications—the material things our culture creates to feed its spiritual emptiness.

The trouble with substitute material gratifications is that we never get enough of what we do not need. In addition, we often do not even know what we are looking for and what we are trying to fill. In our frenzied consumerism to fill our empty spiritual wells, we are blind to the very foundation of our humanness—the qualitative spiritual values of such things as love, trust, respect, and meaningful relationship.

Finally, we must recognize, understand, and accept that our communities are microcosms of the larger society. Thus, at the heart of any serious effort to change how communities function lies the need to address the basic dysfunction of our expansionist economic world view in the larger culture and vice versa.

The Transition

Having been long steeped in the expansionist economic world view, it is too easy to dismiss as impractical idealism any attempt to refocus from political issues to ideas and processes. Yet our world view is counter to the law of entropy, which says that the current theory of social development, our social trance, is seriously flawed relative to the "real world," a notion repeatedly demonstrated by history and repeatedly ignored by society. People therefore fail to see how their accustomed way of life is based on unviable thoughts rooted deeply within our present-day social trance.

For a new world view to emerge, says Taylor,[26] its creators must be willing to take the time to identify, clarify, and evaluate the underlying assumptions of the existing dominant world view. Failure to do so ensures that the power of habit and the tyranny of the familiar values of the old model will co-opt those of the fledgling view.

Taylor goes on to contrast the current expansionist economic world view with an emerging ecological world view. It seems clear to me, both from my own experience and from Taylor's contrast, that just as the expansionist economic view excludes, to the greatest extent possible, the ecological lessons of history and current scientific knowledge, the ecological view excludes, to the greatest extent possible, economics in the social sense.

Both views have thus become what they are against—being exclusionary. This is ironic because ecology and economics have the same root: *eco* from the Greek *oikos*, a house. Ecology is the knowledge of the house, and economics is the management of the house, and it is the same house! Both views thus ignore the health, vitality, and sustainability of the whole.

Today's clutching at cheap material substitutes for spiritual truth can satisfy only temporarily before setting up the need for more and more cheap material substitutes, the pursuit of which squanders the world's resources. Is there a way out of this morass? The answer is yes. The way out is a world view that brings the material and the spiritual into harmony, heals the mind/body and human/Nature schism, while simultaneously creating a mutualistic symbiosis between ecology and economics.

Such a view must recognize the value of relationship and accept that the only way anything can exist is encompassed in its interdependent and interactive relationship to everything else, which means that an isolated fragment or an independent variable exists only on paper. As such, every relationship is dynamic, constantly adjusting itself to fit precisely into all other relationships, which consequently are adjusting themselves to fit precisely into every other relationship, and so on *ad infinitum.*

Can so fluid a notion as ever-adjusting relationships be made to work within a rigid, mechanistically oriented, reductionistic social construct powered by the notion of ever-increasing economic expansion? The answer is no, because, through self-reinforcing behavioral feedback loops, our present social paradigm condemns change as a condition to be avoided at almost any cost.

While no one is safe from the many environmental problems engendered by our current world view, which threatens our health and that of the planet, we are not all endangered equally. In addition, the perceived security we have long sought through ever-increasing consumption, militaristic technology, and domination over Nature has actually threatened our long-term social survival. This threat prompted Czech President Václav Havel to observe: "Without a global revolution in human consciousness a more humane society will not be possible."

We must, through a global revolution in human consciousness, change our world view (from one based strictly on a linear commodity-oriented thought process) to a systems approach, where the indicators of both ecological and economic health are rooted in the quality of the relationships between and among the parts. Such a view does not mean that the linear commodity-oriented thought process is summarily discarded. Rather, it means that the linear commodity-oriented thought process is moderated and ultimately constrained by those ecological limitations that control the biological sustainability of social/environmental harmony.

If we have the moral courage and the political will to adopt and implement the concept of sustainability as a component of social evolution in which change is accepted as a process to be embraced, then the notion of ever-adjusting relationships becomes the creative energy that guides a vibrant, adaptable, ever-renewing society through the present toward the future. And because sustainability honors the integrity of both society and its environment, the outcome is a unified world view in which the function of a system defines the system. That is to say that the function defines the composition, which in turn defines the structure, and it is by the visible structure that we tend to characterize a system.

This is all well and good, but "transition" indicates a change. How do we, who seem to be so stuck in our current world view, go about changing? I think we need a lesson from Zen.[27]

A symptomatic person, by definition, keeps repeating the behavior that causes distress while protesting that he or she would rather not be repeating the behavior but cannot help it. This kind of fixation is assumed in Zen to prevent a person from fully experiencing the present moment, which is one way of viewing enlightenment.

The term symptomatic suggests that there is some flaw within the system that causes the manifestation of a visible symptom. The relevance of this to Zen is the basic premise of Zen, which is that human beings are trapped on the revolving wheel of life and keep repeating distressing behavior. The more a person attempts to escape this destiny, the more she or he is caught up in it, because any attempt to change only causes the system to resist and to continue.

One goal of Zen is to free the person from the repeating cycle of the system so that new, spontaneous behavior can occur. It is the assumption of Zen that an attempt to help someone wanting to change or to initiate change oneself will cause a reaction that prevents both the help and the change from occurring. As any element, such as a person within an agency, moves toward change, a reaction within the self-maintaining system counters the movement and prevents the change. How to change without activating the forces resisting it is the paradox of Zen.

Change, in Zen, is described as sudden and discontinuous, rather than as cumulative, in a stepwise fashion. The student of Zen is not taught how to become enlightened; instead, one discovers that one must give up theories about enlightenment to achieve it.

To free a person from the limitations of theories, a Zen master uses the absurd, such as riddles, koans, and actions, to alter a system of classification based on logic—the language of the mind. Instead of relying on the intellect as the moral basis of our decisions, regardless of how logical they may seem, we must learn to recultivate the language of the heart, which is the language of the moment. We must relearn to do what is morally appropriate (other-centered) within our own hearts, in the present of the moment, and have faith that what we send around to others will in the future come around to us.

Unified Systemic World View

The unified systemic world view is predicated on the notion of holism, in which reality consists of organic and unified wholes that are greater than the simple sum of their parts. The following are basic assumptions on which the unified systemic world view is founded: (1) everything exists in relationship to everything else; (2) every relationship is dynamic, constantly adjusting itself to fit precisely into all other relationships, which consequently are adjusting themselves *ad infinitum*; (3) all relationships, including nonmonetary ones, have value; (4) everything, including humans and nonhumans, is interconnected, interdependent, and interactive; (5) all relationships are systems supporting systems supporting systems *ad infinitum*; (6) the whole is functionally greater than the sum of its parts; (7) processes have primacy over components; (8) a system is defined by how it functions, not by the shape, number, or arrangement of its component parts; (9) the integrity of the environment and its ecological processes has primacy over human desires when such desires would destroy the system's integrity for future generations; (10) Nature determines the limitations of human endeavors; and (11) the relevancy of knowledge depends on its context.

In the unified systemic world view, the behavior of a system depends on how individual parts *interact* as functional components of the whole, as opposed to what each isolated part is doing, because the whole is understood through the relation/interaction of its parts. To understand a system, therefore, we need to understand how it fits into the larger system of which it is a part. This gives us the above-mentioned view of systems supporting systems supporting systems *ad infinitum.* We thus move from the primacy of the parts to the primacy of the whole, from insistence on absolute knowledge as truth to the relatively coherent interpretations of constantly changing knowledge, from isolated self to self in community, and from attempting to solve old problems to creating new concepts.

In a unified systemic world view, the disenfranchised as well as future generations have constitutional rights that must be accounted for and honored in present decisions and actions. In addition, individual people—as well as their relationships among one another, Nature, and their communities—both have value and are valued, as are all living beings. It is this combined sense of feeling valued as a person and feeling the personal value of contributing positively to one's community that makes sustainability a workable proposition for the present and into the future.

Sixth Element: Accepting Our Ignorance and Trusting Our Intuition, While Doubting Our Knowledge

There is much to learn about the Earth, and there are many good reasons to believe that its complexities are permanently beyond our comprehension. Thus the salient point is not our knowledge but our ignorance, because only our ignorance can be proven. The credibility of our knowledge rests, albeit tenuously, before the jury of tomorrow!

For the present, however, it is necessary to keep the search for truth on its own credible track. This requires that we recognize our tendency not only to form a single hypothesis but also to become so attached to it that any criticism of or challenge to our methods raises our ego defenses. Then comes the massaging (as I've often heard it called in government agencies) of the theory to fit the data and of the data to fit the theory.

In addition, we tend to become "method oriented" rather than "problem oriented" in our thinking and therefore in many of the questions we ask. It is important to recognize that we become method oriented in our questions, because we tend to think that through our experiments—our methods—we are learning the Truth about Nature when in fact we are learning only about our experimental designs— again, our methods—and our assumptions and expectations.

Margaret Shannon, discussing what she calls "feminist theory," makes some important observations, which are pertinent here.[28] She says that one of the primary assumptions of feminist theory is that the realities we accept as obvious, neutral, objective, and simply the way the world works are actually structures of power. They are created as we think and live. They are created by our rendition of history, our understanding of ourselves, our society, and our world—and they are always partial with respect to the whole.

She goes on to say that it is exactly when a problem seems the most objective, neutral, and empirically based that the viewer is most blinded to the power of the bias shaping his or her perspective and is most apt to minimize other forms of knowledge and deny subjectivity. It is precisely those times when things seem most objective, neutral, and empirical that they are most likely to be oppressive.

It is thus impossible to accurately "represent" Nature through science, because scientific knowledge not only is a socially negotiated, rigid construct but also is a product of the personal lens through which a scientist peers. To which author Manly P. Hall adds: "As a great part of so-called knowledge is merely accredited opinion...the intellectualist may be over-well versed in fallacies....The intellectual man, limited in his learning, is helpless in the presence of the unknown."

Although scientists may attempt to detach themselves from Nature and to become "objective," they are never completely successful. They are part of Nature and must participate with Nature in order to study Nature. Nevertheless, the role of science is to help humanity understand the physical/biological principles governing the Universe and how we are related to them as an inseparable, functioning part of that Universe.

In addition, every scientist sees through his or her lens but dimly, first because we cannot detach ourselves from Nature and second because all we can judge as fact are our own perceptions, which are always colored by our personal lenses. Even history, which we tend to think of as fact, is viewed individually through interpretation of that which we perceive. "My view of history," says Arnold Toynbee, "is itself a tiny piece of history; and this mainly other people's history and not my own...."[29]

The irony of scientific research is that nothing can be proven; it can only be disproven. Therefore, we can never "know" anything in terms of knowledge. We can "know" only in terms of intuition, which is the knowing beyond knowledge and not admissible as evidence in modern science, which is part of the oppressiveness to which Margaret Shannon referred. Whatever truth is, it can only be intuited and approached, never caught and pinned down.

Knowledge, therefore, which is external to a person, is not "knowable," and intuition, which is internal to a person, is not knowledge and therefore is not subject to disproof. Intuition is inner sight—individualized, inner knowing that is beyond knowledge—for which proof is unnecessary and explanation impossible.

Intuition, a mode of knowing widely accepted since ancient times, has been clouded with ambiguity and controversy since the advent of the reductionistic mechanical mindset, which swept Western society more than a century ago. For some, intuition is merely a meaningless by-product of unconscious processes, but for others it is a harbinger of the deepest truths. Intuition, an instantaneous, direct grasping of reality, is the source of our deepest truths, that sense of unquestionable knowing, of which even John Stuart Mill, a pillar of the empirical method, said: "The truths known by intuition are the original premises from which all others are inferred."[30]

Intuition is therefore more than a guide to truth, says sociologist Pitirim A. Sorokin. It seems to be the foundation of our conception of the beautiful and the good, and it is well understood in philosophy that our aesthetic and moral judgments are based on deep subjective feelings.[30]

It is also well known that there are, among the founders of quantum physics, a number of self-avowed mystics and intuitionists, such as Niels Bohr, Albert Einstein, Sir Arthur Eddington, and Eugene Wigner. To them, such principles as time, space, and the conservation of forces are deep intuitions, if not acts of faith.[30]

Knowledge, on the other hand, is a collective outer experience of humanity and society's subjective judgments about things, which is shareable, but only relative, and disprovable—the social construct of science. The truth of any piece of knowledge cannot be known and is therefore explainable only in the illusions of its appearance.

Thus the actual objects of our inquiries, the formulations of our questions and definitions, and the mythic structures of our scientific theories and facts are social constructs. Every aspect of our scientific theories, facts, and practices—including "scientific method" itself—are but expressions of contemporary socio/political/economic interests, cultural themes and metaphors, personal biases, and personal/professional negotiations for the power to control, albeit momentarily and minutely, the scientific knowledge of the world.

Facts that scientists construe to be statistically true statements about Nature are demonstrated to be concrete, deified, magic outcomes of the social process of fabricating statements about the world so as to distinguish order from chaos. Thus, instead of scientific consensus being achieved when the "facts" reach the state of "speaking for themselves," scientists come to a consensus when the political, professional, and economic costs of refuting them make further negotiation untenable.[31]

There is, however, no single reality, but rather a multiplicity of realities, the representation of which depends on one's position in the process of negotiating an acceptable social view of reality. Thus, regardless of the question, the reproducibility of the experimental design and methods does not mean that the results represent anything about Nature. The reproducibility of the experimental design shows merely that a particular negotiation of reality is reproducible under a certain set of conditions. Thus, the results of every experiment may be valid, if unprovable, only because the experimental design tells us nothing about the results. It tells us only that the reproducibility of the experimental design is socially acceptable according to a consensus of scientific opinions.

Science, policy, and organization therefore represent a structure of attention that recognizes some qualities, features, and relationships at the expense of others. The tendency in science, all too frequently and uncritically, is to use abstract models as concrete examples based on postulates about how a part of the world works by claiming that the models portray only that which is "really there."[32]

The authors of theories and models say that the weakness of their theories and models is that they are only partial, not that they can be questioned about their fidelity to what might really be happening in the world. While this claim may be questioned and greater humility sought regarding scientific knowledge, scientific language reveals the arrogance of science as the namer of that which is to be "real." In this way, science becomes an ideology that best serves the interests of science and specialized expertise in a technological society.[32]

The problem is that we confuse the limiting nature of scientific method with the nature of ordinary experience. While scientific method tends to be reductionistic and mechanical, life experiences are synthetic holograms made of complex threads of past experiences, present perceptions, dreams of future opportunities, and fears of future circumstances. Life is thus a moment-by-moment kaleidoscope of a synthetic present in eternal flux within and between internal and external realities.

If, therefore, we are going to ask intelligent questions about the future of the Earth and our place in the scheme of things, we must understand and accept that most of the questions we ask deal with cultural values, which cannot be answered through scientific investigation. Nevertheless, scientific investigation can help elucidate the outcome of decisions based on these values and must be so employed. We also must be free of scientific opinions based on "acceptable" interpretations of scientific knowledge. In addition, we would be wise to consider the gift of Zen and approach life with a beginner's mind—a mind simply open to the wonders and mysteries of the Universe.

A beginner sees only what the answers might be and knows not what they should be. If, on the other hand, I become an expert, I think I know what the answers should be and can no longer see what they might be. The beginner is free to explore and to discover a multiplicity of realities, while the expert grows rigid in a self-created prison of a single pet reality, which often turns into an obsession to be protected at any cost. The beginner thus understands the question better than the expert does.

If we are going to ask intelligent questions, we must be open to multiple hypotheses and explanations, and we must be willing to accept a challenge to our ideas in the spirit of learning, rather than as an invitation to combat. The greatest triumphs in science are not, after all, triumphs of facts but rather triumphs of new ways of seeing, of thinking, of perceiving, and of asking questions. To achieve sustainability, therefore, we must learn to accept our ignorance and trust our intuition, while doubting our knowledge.

Seventh Element: Specifying What Is to Be Sustained

The meaning of "sustainability" is clear in that it is consistently used, either explicitly or implicitly, to mean a continuity through time. Whatever ambiguities exist are not with the concept of sustainability itself, but rather with the implications of sustainability in a given context. For example, what is the scale of time within which something must exist in order to be considered sustainable or sustained? This question does not impinge on the meaning of sustainability itself as a mark of continuity; it is only a matter of duration or scale.

To sustain anything, one must specify what is to be sustained and what is not. This is an important concept, because when one elects to sustain one thing, one simultaneously elects not to sustain something else.

For example, the language of the Multiple Use Sustained Yield Act of 1960, although of good intent, is based on an economic assumption, which is totally at odds with ecological reality. The assumption is that biological processes in a forest ecosystem remain constant while we humans strive to maximize whatever product or amenity of the forest seems desirable.

Sustained yield is thus an economic concept in which the volume of wood fiber to be cut annually is predetermined by the economic targets of industry. These targets have nothing to do with the ecological capability of the forest to produce that volume on a sustainable basis without impairing its ability to function ecologically over time. "Sustained yield" signifies that the Congress of the United States has chosen to sustain the *yield* rather than the biological processes of the forest that produces the yield.

It works this way: Ninety percent of the timber cut in what is now the Willamette National Forest of western Oregon during the first three decades of this century occurred below 4,000 feet elevation. In addition, every increase in the technology of logging and the utilization of wood fiber has speeded the cutting of forests.

From 1935 through 1980, there was a geometric increase of 4.7 percent per year in the annual volume of timber cut in the Willamette National Forest, which means a doubling of volume cut every 15 years. By the 1970s, 65 percent of the timber (yield) was cut in forests above 4,000 feet elevation, where the average tree harvested was not only progressively younger and smaller than that of low elevation but also slower growing in a much more biologically fragile forest. To sustain the same *yield* (=volume of wood fiber cut) as that which came from the most biologically resilient and productive forests in Oregon (those below 4,000 feet elevation), it was necessary to increase the annual acreage cut in the less productive high-elevation forests (those above 4,000 feet) by five times.[33]

If, however, the real intent of Congress had been to maintain the productivity of forests, consider the following question, which represents a fundamental shift in perception: How would we be thinking about and acting toward our forests, especially our national forests, if the Multiple Use Sustained Yield Act had been conceived of and written as the Multiple Use Sustain*able* Yield Act or even as the Multiple Use *Sustainable Forest* Act?

In the latter two cases, that which would have been selected to be maintained in a sustainable condition would have been forest processes and long-term productivity of the soil. This choice would have protected the system's biological sustainability and thereby ensured sustainable commodity production, rather than ensuring short-term economic commodity output at a predetermined level at the cost of the system's biological sustainability.

While sustainability does not exclude the extractive use of resources, it does demand a balanced approach to their long-term maintenance, extraction, and use. This means that the economic divesting of resources from any ecosystem must be at least balanced by the biological reinvestment of resources in that same system, regardless of the economic impact on the profit margin—something that is not now happening. The concept of balance is important because we live between two spheres—the atmosphere (air) and the lithosphere (soil)—and if we destroy either one, we will be the authors of our own extinction.

If, for example, we imagine sustaining our current expansionist approach to economics into the future, we soon bump into environmental crises and come up hard against the need to reframe the old economic paradigm, namely, that continued growth (often called development) can solve all our social problems.

We now assume, in our strategies to raise material prosperity, that ever-expanding urban/industrial growth and more technological development are necessarily and ethically good because they result in the production of greater capital gains with which to purchase more material goods, thus making life "better" than it presently is. But if the importance of such growth and development is only to allow us to achieve ever-higher levels of material prosperity, is sustaining the accompanying environmental degradation socially responsible or ethical? After all, our choice of lifestyle affects the options not only throughout our own lives but also throughout the lives of our children and theirs.

Eighth Element: Understanding and Accepting That Sustainability Is a Continual Process, Not a Fixed End Point

The pivot around which the notion of sustainable development revolves is the recognition that sustainability is a continual process centered in conscious, purposeful moral transformation (both personal and social), as opposed to a condition that once achieved becomes static and thus can be ignored. The outcome of such transformation must be a self-reinforcing feedback loop of ever-growing social/environmental harmony.

Because of the dynamic nature of the evolving ecosystems we attempt to manage, such as a forest, grassland, or ocean, we can only attempt to manage an ecosystem for its evolution and not for a sustained yield of products, which is an economic absolute. Therefore, the only sustainability for which we can attempt to manage is that which ensures an ecosystem's ability to adapt to changing conditions, such as warming of the global climate.

In addition to Nature's disturbances through which ecosystems evolve, ecosystems are also influenced by human-introduced disturbances. In the early years of the nineteenth century, for example, there was an emerging view that there were strong interdependencies among the climate, plants and animals, and the soil, which led to the long-term stability of terrestrial ecosystems in the context of landscapes. But this notion was considered only in the context of a stable climate in the scale of space. Today, a revised concept is emerging, which might be stated as follows: The spatial patterns of ecosystems, including those of forests or grasslands, observed on landscapes result from complex interactions among physical, biological, and social forces over time.[34]

Because most landscapes have been influenced by the cultural patterns of human use, the resulting landscape is an ever-changing mosaic of unmanipulated and manipulated patches of forestland, shrubland, grassland, and other habitats, which vary in size, shape, and arrangement. This spatial patterning is a unique phenomenon that emerges at the level of the landscape in time and changes over time.

With respect to human-introduced disturbances, such as fragmentation of habitat, biogeographical studies have shown that the connectivity of the areas within a landscape is important to the persistence of ecological processes as well as plants and animals in viable numbers in their respective habitats. In this sense, the landscape can be considered as a mosaic of interconnected patches of habitat, such as forested riparian areas that act as corridors or routes of travel between patches of upland forest or other suitable habitat.

To allow and to help ecosystems, such as forests, to migrate in the face of global climate change, we must emphasize diversity in all of its aspects, which in turn will do much toward ensuring the resilience and sustainability of those systems. We also must pay close attention to the patterns we create on the landscape, because the overall connectivity of those patterns will either allow or inhibit the ability of species, biological communities, or even whole ecosystems to migrate with changing climatic conditions.

The ability of a species, biological community, or even whole ecosystem to migrate will determine how much of the genetic variability—adaptability—is lost from the gene pool with the extinction of local populations of plants and animals. Such extinctions inevitably accompany the fragmentation of ecosystems within landscapes and across biogeographical regions. Part of the reason these extinctions have thus far been inevitable is that while people think saving species is important, they have little or no understanding of the cause and effect wrought through the aggregate of human activities on the health of habitats and the environment as a whole.

What decisions we make are up to us, but one thing is clear: While the current trend toward homogenizing ecosystems may make sense with respect to maximizing short-term profits, it bodes ill for the long-term ability of both ecosystems and landscapes to adapt to changing environmental conditions on a sustainable basis. I suggest, therefore, that while we must strive in the present to manipulate ecosystems on a biologically sustainable basis, we must simultaneously recognize that our concept of sustainability may prove to be relatively shorter lived than we anticipated. I say this because the climate, which has long been taken for granted as a constant and thus omitted from economic models, is changing.

As evidence of global warming grows, it becomes increasingly apparent that climate must now be accepted and accounted for as a variable of great uncertainty. We must therefore look beyond the sustainability of ecosystems as isolated entities in time and space to their long-term sustainability as contextual components of landscapes, whose adaptability to changing environmental conditions must be protected.

The long-term sustainability of ecosystems may well depend on the connectivity of the patterns we create on the landscape and the adaptability of those patterns to the forced migration of ecosystems, or components thereof, over time as demanded by changing environmental conditions. This means that we must rise to the challenge of manipulating ecosystems first and foremost for the sustainability of their internal processes, second as adaptable patterns on rapidly changing landscapes, which may shortly be subjected to unprecedented global warming, and third for the goods and services they provide to humanity.

With the above priorities in mind, our responsibility in making current decisions about the sustainability of the adaptive processes of the patterns we create across landscapes is to be mindful of the consequences our decisions will have on the generations of the future and their ability to respond to the conditions that our decisions will create. This is particularly important, because economic and ecological systems are perceived to operate on different scales of time, which means that the long-term detrimental effects of decisions made in favor of short-term profits are ignored. As a result, the long-term, nonmarket value of Nature's services is grossly underestimated.[35]

Considering our present trajectory, where we as a species will ultimately end up, with an environment compatible to our existence or hostile to our existence, depends on the level of consciousness we bring to the questions we ask. And before we can get fundamentally new answers, we must be willing to risk asking fundamentally new questions.

We must therefore look long and hard at where we are headed with respect to the patterns we create on the landscape as we move from the most natural end of the landscape continuum toward the most cultural end. This is important, because the old questions and the old answers have both gotten us to where we are today in the fragmentation of ecosystems and are guiding us to where we will be tomorrow in accelerating such fragmentation toward the ultimate collapse of ecosystems.

We can, however, manipulate our landscapes so that we consciously protect, maintain, and create as many options as possible for both ourselves and the generations to come. To do this, we must understand that the quality of the question is all important.

What is out there? What do we want out there? What are our values? Is what we want ecologically compatible with the landscape as it has evolved over the millennia? What question must we ask to ensure that we maintain as many ecologically viable options as possible, both now and for the future? How must we behave in order to live compatibly with the land? These types of questions must be asked continually and simultaneously with our decisions about what and how to manipulate an ecosystem, and each question must be answered to the absolute best of our ability, even as we redesign the landscape in an attempt to suit our cultural desires.

We also must understand that conflicts over values, either ecological or human, are not battles over numbers but rather battles over different visions of the world order and their respective desirability and long-term ecological sustainability. We cannot, after all, legislate feelings or values, only behavior. In the end, it is the sum of the consciousness of our daily choices of behavior that will make the difference, choices that ultimately are reflected in the questions we ask.

Thus, by its very nature, the sustainability of a dynamic system is a moving target. Although it may be more clearly sighted with increasing spiritual development and greater scientific/social knowledge, sustainability will never be a finite end point in an ever-evolving world.

Ninth Element: Understanding, Accepting, and Being Accountable for Intergenerational Equity

Intergenerational equity means the responsibility of the current generation to its own members and to its descendants. The concept of intergenerational equity or environmental justice, from the human point of view, by its very nature asserts that we owe something to every other person sharing the planet with us, both those present and those yet unborn. "But what exactly," you may ask, "do we have to give?"

The only things of value we have to give are our love, trust, respect, and the benefit of our experiences. These are the essence of human values that we either extend to one another or withhold. These are the values embodied in each and every option we pass forward or withhold in each and every option we foreclose. When everything is said and done, all we have to give the children of today, tomorrow, and beyond is options and some things of value from which to choose.

Tenth Element: Understanding, Accepting, and Being Accountable for Ecological Limitations to Land Ownership and the Rights of Private Property

Private ownership of land is a very recent concept. For example, the pygmies of central Africa, the most ancient of all forest dwellers, hold no enforceable claims to the forests they have inhabited for at least 40,000 years.[36] Indigenous peoples on every continent find the notion of private ownership of land to be both ludicrous and impossible.

How can an individual human being own something that he or she has not created and therefore cannot control? How can an individual own something that has been around for millennia before he or she was born and will continue for millennia after he or she is dead? How can an individual own something that is so obviously part of the global commons in both time and space that it belongs to every living creature in its turn and so to no one individual in particular at any given time? How is it that we in the United States *cannot* legally steal from our adult neighbors in the present (=in space), but we *can* legally steal from our children neighbors in the future (=in time) through nonsustainable overexploitation of resources? (For a discussion of resource overexploitation, see *Resolving Environmental Conflict.*[25])

Society is divided on this issue. Society must therefore decide. Either this peculiarly human notion of land ownership and the rights of private property will continue or people must accept an alternative, such as that of custodial trustee of the piece of Earth one inhabits or otherwise has deed and title to as a living trust for the beneficiaries of the future. It is a question of self-centeredness versus other-centeredness, which means that the morality of the idea of land ownership and unlimited rights of private property must be opened to rigorous debate. And make no mistake, the question of land ownership and the rights of private property is, at its very root, a moral one.

In such a debate, the questions must be: Does the holder of a deed to land of any kind "own" the land in an absolute sense, or is he or she only a custodial trustee thereof? Does such a holder of a deed have the moral right to degrade the productive capacity of the land before passing it on to the next person who must use it? Put a little differently: Does any person have the moral right to steal options from the future for immediate personal gain by irreparably degrading the productive capacity of the land? If not, how can one be granted the legal right to do so?

If the outcome of such debate is in favor of the status quo, then biological sustainability is at best an academic question—and so is every other kind of sustainability. Before sustainability can be tenable, the ownership of land and the unlimited rights of private property must be modified. Such modification must be in the tenor of a person's privilege to enjoy being the custodial trustee of the piece of Earth he or she inhabits or otherwise has deed and title to as a living trust for the beneficiaries of the future. Only then is the biological sustainability of the Earth possible.

Should the status quo prevail, it will do so because of what professor David Orr calls "conservatives against conservation."[37] The following discussion is taken from his insightful article.

"The philosophy of conservatism has swept the political field virtually everywhere," says Orr, and virtually everywhere conservatives have forgotten what conservatism really means. Orr goes on to say that conservative philosopher Russell Kirk proposes six "first principles" of conservatism, based on his "love of order," for which true conservatives are accountable:

1. Believe in a transcendent moral order;
2. Prefer social continuity (i.e. "the devil they know to the devil they don't know");
3. Believe in "the wisdom of our ancestors";
4. Are guided by prudence;
5. "Feel affection for the proliferating intricacy of long-established social institutions";
6. Believe that "human nature suffers irremediably from certain faults."

Nevertheless, it is eighteenth-century British philosopher and statesman Edmund Burke who is considered to be the founding father of modern conservatism and, according to Orr, is "as much admired as he is unread." To Burke, the goal of order is to harmonize the distant past with the distant future through the present, which is the nexus.

Like current Republicans, Burke thought in terms of a contract. But unlike the prevailing Republican contract, which is self-centered for a minority in the present, Burke's contract is between "those who are living, those who are dead, and those who are to be born." Those "possessing any portion of power," says Burke, "ought to be strongly and awfully impressed with an idea that they are in trust." In Burke's contract, freedom is "that state of things in which liberty is secured by the equality of restraint," and not one in which "every man was to regulate the whole of his conduct by his own will."

As the ecological shadow of the present stretches increasingly over the generations of the future, the wisdom of Burke's concern for the justice and welfare of the generations yet unborn becomes more evident. If conservatism means anything at all, says Orr, it means the conservation of what Burke called "an entailed inheritance derived to us from our forefathers, and to be transmitted to our posterity; as an estate belonging to the people." It *did not* mean preserving those rules by which one class enriches itself at the expense of another.

"What is conservative," asks Orr, "about squandering for all time our biological heritage under the pretext of protecting temporary property rights?" Present-day conservatives scorn efforts by the public to protect such things as endangered habitats (like old-growth forests), endangered species (like spotted owls, coho salmon, and red wolves), clean air, and clean water. Almost any restriction placed on the rights of an individual to use land as private property is being viewed increasingly as an unlawful "taking," even when such use would irreparably damage the land and its surrounding environment. How, one might ask, is it any more of a *lawful* "taking" when one degrades land in the present that must be used in an impoverished condition by someone in the next generation and beyond?

Even John Locke, from whom we have derived much of our land-use law and philosophy, said that "nothing was made by God for Man to spoil or destroy." "The point," says Orr, "is that John Locke did not regard property rights as absolute even in a world with a total population of less than one billion, and neither should we in a world of 5.7 billion."

"What," asks Orr, "is conservative about conservatives' denial of the mounting scientific evidence of impending climatic change?" Climate change will have rapid, self-reinforcing feedback loops that could change the nature of the Earth's hospitality to human life for all time. What right do we have to run such a risk, when the consequences belong to the generations of the future and they are not here to participate in the choice?

What, asks Orr, is conservative about perpetual economic expansion when it not only has changed the Earth more radically than any other force in modern times but also is rapidly destroying communities, traditions, cultural diversity, and whole ecosystems throughout the world? What is conservative about passing forward a despoiled legacy to the future?

Social/environmental sustainability requires no less than the first of Russell Kirk's "first principles": that humanity must be grounded in the belief in a transcendent moral order in which we humans, as trustees for future generations, are accountable to a Higher Authority. Anything less is not sustainable!

Edmund Burke put the capstone on land ownership and the unlimited rights of private property as a sustainable proposition when he wrote:

> Men are qualified for civil liberty in exact proportion to their disposition to put moral chains upon their own appetites....Society cannot exist unless a controlling power upon will and appetite be placed somewhere, and the less of it there is within, the more there must be without. It is ordained in the eternal constitution of things that men of intemperate minds cannot be free. Their passions forge their fetters.[37]

The Human Relationships

Everything must exist in relationship to something else; conversely, nothing can exist out of relationship to something else. Relationship, the quintessential thread running through time and space, is the essence of sustainability. It is also the essence of humanity because we are creatures who must share to find value, regardless of how it is defined.

For example, what is your favorite place on Earth? How would you feel if you owned it outright? How would you feel if you owned it outright but were the last person on Earth and had no one with whom to share the experience? How would you feel if you owned your favorite place outright and had one other person with whom to share it? Two other people? Three other people?

I once asked this question of an audience, and a woman told me that she would not want to share her favorite place with anyone because they would just ruin it. But notice that she still spoke of her favorite place in terms of relationship with someone else. How we choose to respond to someone else is thus our determination of how we choose to share our emotions, knowledge, and experience of our self-concept with the other person. The other person in turn must make the same determination about us.

Mahatma Gandhi, in a conversation with his grandson, Arun, a few weeks before his assassination, handed him a talisman upon which he had engraved "Seven Blunders." Out of these seven blunders, said Gandhi, grows the violence that plagues the world. The blunders are:

Wealth without work
Pleasure without conscience
Knowledge without character
Commerce without morality
Science [and technology] without humanity
Worship without sacrifice
Politics without principles

Gandhi called these imbalances "passive violence" and said that "We could work till doomsday to achieve peace and would get nowhere as long as we ignored passive violence in our world."

To his grandfather's list of seven blunders, Arun added an eighth:

Rights without responsibility

That, says Donella Meadows, was in 1947, almost 50 years ago, and today every blunder remains institutionalized, built into our corporations, our government, and our very culture. In fact, she says, we actively practice them and even take pride in some. She then goes on to add more blunders:

Justice without mercy
Order without freedom
Talking without listening
Stability without change
Private interests without public interests
Liberty without equality
Or, in every case, vice versa

If, as Meadows points out, you suggest putting morality into commerce, you are accused of being against commerce. Question wealth without work and you are a jealous hater of rich people or out to undermine capitalism, says Meadows.[38] Whenever we make our private or public discussions about relationships an either/or proposition, we lock ourselves into an unwinnable position of defending a simplistic point of view that has little or nothing to do with the larger reality beyond that of our own perception.

Relationships are the strands in the web of life, and there is no escaping the web. For this reason, the following human relationships are central to sustainability: (1) *intra*personal, (2) *inter*personal, (3) between people and the environment, and (4) between people in the present and those of the future.

Intrapersonal

An intrapersonal relationship is the relationship that exists within a person. It is the individual's sense of his or her own spirituality, self-worth, personal growth, and so on. In short, it is what makes that person conscious of and accountable for his or her own behavior and its consequences. The more spiritually conscious one is, the more other-centered one is, the more self-controlled one's behavior is, and the greater one's willingness to be personally accountable for the outcome of one's behavior with respect to the welfare of fellow citizens, present and future, and the Earth as a whole.

There is an interesting correlation between self-centered and other-centered people when it comes to how they see themselves fitting into their environment. Self-centered people tend to be *piece thinkers*, who focus on individual pieces of a system or its perceived products in isolation of the system itself. Other-centered people tend to be *systems thinkers*, who focus on the welfare of the system as a whole.

A piece thinker is oriented to seeing only the desirable pieces of a system (be they economic or environmental) and seldom accepts that removing or isolating the perceived desirable piece (or eliminating an undesirable one) can or will negatively affect the productive capacity of the system as whole. This person's response typically is: "Show me; I'll believe it when I see it."

In contrast, a systems thinker sees the whole in each piece and is therefore concerned about tinkering willy-nilly with the pieces because he or she knows such tinkering might inadvertently upset the desirable function of the system as a whole. A systems thinker is also likely to see himself or herself as an inseparable part of the system, whereas a piece thinker normally sets himself or herself apart from and above the system. A systems thinker, being other-centered, is willing to focus on transcending the issue in whatever way is necessary to frame a vision for the good of the future, whereas a piece thinker is self-centered and focuses only on what he or she perceives as personally beneficial.

The more of a piece thinker a person is, the more reticent that person is to change, and the more readily he or she denies the effect of his or her behavior on the system as a whole. This type of individual sees change as a condition to be avoided because he or she feels a greater sense of security in the known elements of the status quo, especially when money is involved. Conversely, the more of a systems thinker a person is, the more likely he or she is to risk change on the strength of its unseen possibilities and its potential good for others.

A piece thinker is likely to be very much concerned with land ownership and property rights and wants as much free rein as possible to do as he or she pleases on his or her property, often without regard for the consequences to future generations. The more of a piece thinker a person is, the greater the tendency to place primacy on people of one's own race, creed, or religion, as well as on one's own personal needs, however they are perceived. The more of a piece thinker a person is, the greater the tendency to disregard other races, creeds, or religions, as well as nonhumans and the sustainable capacity of the land. Also, the more of a piece thinker a person is, the more "either/or" one's thinking tends to be.

A systems thinker, on the other hand, is likely to be concerned about the welfare of others, including those of the future and often their nonhuman counterparts, because a systems thinker has a greater sense of how things fit together in a functional whole than a piece thinker does. Systems thinkers also tend to be concerned with the health and welfare of planet Earth in the present for the future. And they more readily accept shades of gray in their thinking than do piece thinkers.

The upshot is that sustainability is possible to the degree to which other-centered systems thinkers hold sway in a community or society, which brings us to interpersonal relationships.

Interpersonal

If someone in a store is rushing blindly to get somewhere and shoves you out of the way, you have a choice in how you respond to being shoved. You can get angry, impatient, and say something nasty, or you can be patient, kind, and understanding. Your thoughts and actions are the seeds you sow each time you make a choice.

We always have a choice, and we must choose. If we do not like the outcome of our choice, we always have the choice of choosing again. We are not, therefore, victims of our circumstances but rather consequential products of our choices. And the more we are able to choose love and peace over fear and violence, the more we gain in wisdom and the more we live in harmony. This is true because what we choose to think about determines how we choose to act, and our thoughts and actions set up self-reinforcing feedback loops or self-fulfilling prophecies that become our individual and collective realities.

It is just such self-reinforcing behavioral feedback loops based on competition for resources that are destroying our environment. As long as competition is the overriding principle of our economic system, we can only destroy our environment because it has become the battle-field in which the war of competition is fought. Our overemphasis on competition in nearly everything fosters the material insecurity that often manifests as greed.

Another tendency of human beings faced with a perceived threat to their sense of material survival is to defend a point of view. There are, however, as many points of view as there are people, and everyone is indeed right from his or her vantage point. Therefore, no resolution is possible when each person is committed only to winning agreement with his or her position. The alternative is to recognize that "right" versus "wrong" is a judgment about human values and is not a winnable argument. It is best, therefore, to define the principle involved in the discussion as the fundamental issue and focus on it. An issue, usually perceived as a crisis, becomes a question to be answered, and in struggling toward the answer, both positive and negative options not only become apparent but also become a choice.

For us in the United States at least, a crisis is too often in our point of view because we tend to perceive the world through a disaster mentality, regardless of evidence to the contrary. We tend to focus on and cling to a view of pending doom, in part because of the emotional discomfort of an unknown future heightened by daily news with graphic portrayals of disasters worldwide.

Fear, a projection into the future of unwanted possibilities, breeds weakness, a state in which there is little time or energy left to develop other areas of life. Out of the weakness of fear, men too easily and too often turn to war in an effort to assert what little power they think remains to them.[39]

Recent years have demonstrated just how mindlessly cruel cultures can be when they live in proximity to and in dread of one another. The psychological and spiritual result of living under such heinous conditions deadens the mind while it savages the heart. Yet the cruelties of cultures and the violence of individuals are the food and drink of the news media.[39]

The instantaneousness of today's news does not give us time to assimilate the stories within the context of global proportions. News came more slowly in olden times and could more readily be kept in proportion relative to the time and area covered by the news. Today, however, newsworthy disasters all seem to happen instantly in our homes via television and can become so overwhelming that we are emotionally numbed by them even as they augment our fear of our own unknowable future. In addition, insurance companies continually foster a disaster mentality.

Consider that insurance companies are betting, based on calculated probabilities, that nothing will happen to you as they take your money, and that you, by purchasing insurance, are betting blindly that a disaster will befall you in the future. You, therefore, are betting against yourself and your future. And it is just this disaster mentality that causes many frightened people to become increasingly self-centered.

For sustainability to be possible, however, self-centeredness must blend into other-centered teamwork. Setting aside egos and accepting points of view as negotiable differences while striving for the common good over the long term is necessary for teamwork. Unyielding self-centeredness represents a narrowness of thinking that prevents cooperation, coordination, possibility thinking, and the resolution of issues. Teamwork demands the utmost personal discipline of a true democracy, which is the common denominator for lasting success in any social endeavor.

But even if we exercise personal discipline in dealing with current environmental problems, most of us have become so far removed from the land sustaining us that we no longer appreciate it as the embodiment of continuous processes. Attention is focused instead on a chosen product, the success or outcome of management efforts, and anything diverted to a different product is considered a failure. It is time, therefore, to re-evaluate the philosophical underpinnings that gird our notions of Nature, community, and society and how they can sustainably be integrated into a common future.

Between People and the Environment

Sustainability means that development programs must, to the extent possible, integrate the local people's requirements, desires, motivations, and identity in relation to the surrounding landscape. It also means that local people, those responsible for development initiatives and their effect on the immediate environment and the surrounding landscape, must participate equally and fully in all debates and discussions, from the local level to the national.[40] Here, a basic principle is that programs must be founded on local requirements and cultural values in balance with those of the broader outside world.

Some time ago, I attended a meeting on the development of rural communities at which economic diversification was the sole focus of discussion. It soon became apparent that the group had no idea of the importance of landscape to the identity of a community. For example, a logging community is set within a context of forest, a ranching community is set within the context of lands for grazing, and a community of commercial fishers is usually set along a coastline, be it a great lake or an ocean.

The setting of a community helps define the community because people select a community for what it has to offer them within the context of its landscape. The setting therefore helps create many characteristics that are unique to the community. By the same token, the values and development practices of a community alter the characteristics of its surrounding environment.

In addition to the surrounding environment is the constructed environment within a community that is also part of its setting and therefore its identity. This includes the buildings, zoning, design of transportation systems, and the allowance of natural occurrences within the structured setting.

In turn, a community's world view defines its collective values, which determine how it treats its surrounding landscape. As the landscape is altered through wise use or through abuse, so are the community's options altered in like measure. A community and its landscape are thus engaged in a mutual, self-reinforcing feedback loop as the means by which their processes reinforce themselves and one another.

Each community has physical, cultural, and political qualities that make it unique and more or less flexible. The degree of flexibility of these attributes in a community is important because sustainable systems must be ever flexible, adaptable, and creative.[41] The process of sustainable development must therefore remain flexible, because what works in one community may not work in another or may work for different reasons.

Beyond this, the power of sustainable development comes from the local people as they move forward through a process of growing self-realization, self-definition, and self-determination. Such personal growth opens the community to its own evolution within the context of the people's sense of place, as opposed to coercive pressures applied from the outside.

Sustainable development encompasses any process that helps people meet their requirements, from self-worth to food on the table, while simultaneously creating a more ecologically and culturally sustainable and just society for the current generation and those that follow. Due to its flexibility and openness, it is perhaps more capable than other forms of development in creating such outcomes because it integrates the requirements of a local community with those of the immediate environment and surrounding landscape, while instilling a relative balance between the local community and the larger world of which it is a part.

Between People in the Present and Those of the Future

We and our leaders must now address a moral question: Do those living today owe anything to the future? If our answer is no, then we surely are on course, because we are consuming resources and polluting the Earth as if there were no tomorrow. If, on the other hand, the answer is "Yes, we have an obligation to the future," then we must determine what and how much we owe, because our present, nonsustainable course is rapidly destroying the environmental options for generations to come. Meeting this obligation will require a renewed sense of morality—to be other-centered in doing unto those to come as we wish those before us had done unto us.

To change anything, we must reach beyond where we are, beyond where we feel safe. We must dare to move ahead, even if we do not fully understand where we are going, because we will never have perfect knowledge. We must ask innovative, other-centered, future-oriented questions in order to make necessary changes for the better.

True progress toward an ecologically sound environment and an equitable world society will be expensive in both money and effort. The longer we wait, however, the more disastrous becomes the environmental condition and the more expensive and difficult become the necessary social changes.

No biological shortcuts, technological quick fixes, or political promises can mend what is broken. Dramatic, fundamental change, both frightening and painful, is necessary if we are really committed to the children of the world, present and future. It is not a physical question of can we or can't we change, but rather one of human morality—will we or won't we change. If we are to approach social/environmental sustainability while there is still time, we must commit ourselves to do so and then get on with it. The outcome of such a commitment rests on two questions: (1) When is enough enough? and (2) Are the consequences of our decisions reversible?

The Questions

Old questions and the old answers have led us into the mess we are in today and are leading us toward an even greater mess tomorrow. We must therefore understand that the answer to a problem is only as good as the question and the means used to derive an answer.

Before we can get fundamentally new answers we must be willing to risk asking fundamentally new questions. This means that we must look long and hard at where we are headed with respect to the quality of our environment and to the legacy we are leaving the children of the world.

Heretofore we have been more concerned with getting politically right answers than we have been with asking morally right questions. Politically right answers validate our preconceived economic/political desires. Morally right questions would lead us toward a future in which environmental options are left open so that generations to come may define their own ideas of a "quality environment" from an array of possibilities.

Under the influence of the expansionist economic world view, any attempt to refocus from immediate political issues to long-term processes and futuristic ideas is dismissed as impractical idealism. Further compounding the belief that long-term processes and futuristic ideas are merely impractical idealism is the notion of "conversion potential." For many people, the only value of anything is its "conversion potential," which dignifies with a name the concept that Nature, having no intrinsic value, must be converted into money before any value can be assigned to it. All Nature is thus seen only in terms of its conversion potential.

Western society was lulled to sleep by the thinking of men such as Descartes and Newton, but we are gradually awakening to the flawed nature of their premises. The new vision of a single organic whole is being derived through the revolution in physics, primarily quantum mechanics and the work of Albert Einstein et al. But conservative thinking has yet to catch up with the knowledge of modern physics and a changing world view.

Over time, science, as imperfect as it is, has performed a vital function. It has stimulated our imagination, challenged old ways of thinking, explored unknown phenomena, excited our sense of wonder and awe, elucidated relationships, and demonstrated the fuzziness of our world view. Science has shown that the sharp, clear lines of the mechanistic myth were derived by reducing ecological variables to economic constancies. It has shown that the world in which we live is an interactive, interconnected, interdependent system in which the whole is expressed by the function of its parts and not by the parts themselves. And science has shown, albeit subtly, that there is no such thing as an independent variable, an interchangeable part, or a predictable outcome.

The salient point, therefore, is our illusion of definitive knowledge. And it is exactly when we are so certain of our knowledge that we are often so abysmally unaware of our ignorance.

Self-centered (=product-oriented) thinking thus argues to retain the expansionist economic decision making of the old reductionistic mechanical premise. Other-centered (=process-oriented, systems-oriented) thinking argues for an evolving unified systemic world view even though it is only now emerging into our consciousness. The conflict in decision making, therefore, is between self-centered and other-centered values based on different world views, something science can address only indirectly.

The future sustainability of human society lies within the context of a sustainable environment. But politics flows from long-term scientific inquiry back to immediate competing social values. It is these values to which the more self-centered, product-oriented segment of humanity pays the most attention, despite the potentially detrimental effects on environmental sustainability, present and future. The question, therefore, is how to integrate science and sociology, intellect and intuition, spirituality and materialism in such a way that we can have a more holistic perspective of the world in which we live and the future must inherit.

The answer is that we must pay vastly more attention to the questions we ask. A good question, one that may be valid for a century or more, is a bridge of continuity among generations. We may develop a different answer every decade, but the answer does the only thing an answer can do: it brings a greater understanding of the question. An answer cannot exist without a question, and so the answer depends not on the information we derive from the illusion of having answered the question but on the question we ask.

In the final analysis, the questions we ask guide the evolution of humanity and its society, and it is the questions we ask, not the answers we derive, that determine the options we bequeath to the future. Answers are fleeting, here today and gone tomorrow, but questions may be valid for a century or more. Questions are flexible and open ended, whereas answers are rigid, illusionary cul-de-sacs. The future, therefore, is a question to be defined by questions.

With the foregoing in mind, consider again the following questions: (1) When is enough enough? and (2) Are the consequences of our decisions reversible?

When Is Enough Enough?

Once again, when is enough of something enough? This seems like a simple question. When you have eaten your fill, for example, you quit eating because you know you have had enough, for the moment at least. But what about "enoughness" in the material sense, other than being immediately satiated with food?

Although often referred to as a "standard of living," material enoughness really has to do with our sense of survival through competition, which, as mentioned earlier, is usually based on a disaster mentality. "Overtaking," according to journalist Jay Griffiths,[5] "is a cultural emblem." The kick for a company is not to be wealthy, in global financial terms, says Griffiths, but to be "wealthier than its competitors," streamlined to overtake like a race car.

This sensation of overtaking is acted out in our consumerism, where our "drug-like hallucinations of happiness rely on the fact that once needs are met, desires must be aggrandized." The ever-increasing rates of acquisition of unnecessary products and their faster disposal feed the manufacturing industry first and garbage dumps, euphemistically called sanitary landfills, second. In a socially competitive world, where speed and overtaking is an index of status, the poor travel slower than the wealthy, and their time is deemed less valuable.[5]

Recall the earlier discussion of the forest of Čergov, in Slovakia, that had been logged with horses in a biologically sustainable manner for centuries. It had been logged sustainably because the people of the small mountain villages who did the logging had enough of what they needed to satisfy them.

Their centuries-old vibrant communities had been in harmony with the landscape. They had enough forest to produce quality timber to keep the people employed for generations. They had clean water and clean air. They had fertile soil for farming, which was combined in the community with the logging of timber and the milling of lumber, which was sold in the cities. And they passed an intact cultural heritage from generation to generation.

But then the Communists who controlled the Slovakian government decided that they did not have enough fast enough. So they destabilized the centuries-old social/environmental harmony of the mountain villages from the outside. To satisfy their wants, they introduced nonsustainable technology in the name of modern forestry.

Chain saws replaced hand-held crosscut saws; Caterpillar tractors and logging trucks replaced horses; clearcut logging replaced careful selective logging; imported loggers replaced local loggers; distant sawmills replaced local mills; local jobs were replaced by unemployment or long commutes to distant cities; the value of time-honored skills was replaced by a sense of present worthlessness; the continuity of the centuries-old cultural heritage was replaced by an instantaneous loss of cultural identity; economically designed monocultural plantations replaced the diverse forest; erosion replaced soil stability and formation; flooding replaced water infiltration and storage. All this happened within less than a decade because enoughness was replaced by greed.

What about us in the United States? As an admittedly—even *proudly*—materialistic society, we tend to ask ourselves over and over such questions as: What do I want? How much do I want? How much can I get? How fast can I get it? How can I get it for little or nothing?

I remember, for example, two small airlines in Oregon, both of which had the opportunity to be sustainable, but neither of which knew when it had enough. The first of these airlines had an excellent business. It seemed to be booked to capacity during the six or so years that I used it. Then, suddenly, the airline decided that it needed to grow, and so it added a route between Oregon and Idaho, where it had not flown before. That extra route caused the company to go bankrupt within a year.

The second airline started with one plane and began immediately building a business. But instead of waiting until it had at least a sustainable business with its one plane, it added a second plane and was bankrupt within a year. In both cases, these companies could have been sustainable if they had known when to consciously stop growing—when enough was enough.

On the other hand, consider a lumber company in California that owned land with redwood trees on it. The company was in fact sustainable, had been for nearly a century, and planned to be so for centuries to come. It cut and milled its timber at a rate that would allow its planted trees to grow for 300 years before they would be harvested. The company treated it employees as partners and cared for them, and their children, as though they really mattered, which they did.

Then, because the company had become so valuable financially, due in large measure to its long-term sustainability, it became the target of a hostile takeover. The first act of the person who took over the company was to clearcut and liquidate the standing timber to pay off the junk bonds he had used to make the takeover possible, thus destroying the sustainability of the company and therefore its long-term monetary value, along with the job security of its employees.

Finally, there is a wonderful book by Max DePree[42] about the Herman Miller furniture company, which has not only been sustainable for years but also is a fine example of the philosophy and leadership necessary to run a sustainable company. Rather than trying to explain the content of the book, I recommend you read it.

Although sustainability is possible, we in the United States have evolved into the "me, now, I never have enough, I want it for nothing" society. Such social immaturity has posed some interesting problems for the people of land management agencies who feel they are somehow mandated to meet the "public demand" for products. "Meeting the public demand" has long been the timber industry's cry of justification for the unbridled clearcutting of the world's old-growth forests.

Our materialistic societal appetite has reached a compulsive, addictive state in which to want is to have to have! We have made synonyms of desire, want, need, and demand, and in so doing we have lost sight of the ecological reality of both the present and the future. The equating of desire, need, and demand with every itch of "want" becomes clear when considering mail-order catalogs. I did not know I *needed* something until I saw it in the pages of a catalog. Now all of a sudden I just have to have it. I simply cannot live without it. And the catalogs are relentless! (Having said this, I realize that there are many people who wake up scared every morning and go to bed scared every night because they do not, in fact, have enough of the material things of life to feel even reasonably secure.)

Society's present collective "wants, desires, needs, and demands" are rapidly, if they have not already, outstripping the world's capacity to supply them, and this is without even taking into account the "wants, desires, needs, and demands" of the coming generations. Nor does it take into account the incredibly stupid waste of global resources in war—any war for any reason.

If we as consumers continue to insist on feeding our insatiable appetites and destroying incredible amounts of precious resources in war, we will surely destroy the Earth even without the use of nuclear weapons. Greed takes a little longer than war, but the result is the same.

The questions we are asking now are unwise questions because they are based on greed. To date, for example, we ask: What is the absolute maximum that we can get out of Nature? What is the absolute minimum that we must legally leave behind for whatever reason?

If human society is to survive, however, the time has come to ask different questions: How much of any given resource is *necessary* to leave intact in Nature as a biological reinvestment in the health and continued productivity of the ecosystem for the benefit of both our-selves and the generations to come? How much of any given resource is *necessary* for us to use if we are to live in a reasonably comfortable lifestyle?

Necessity, in this sense, is a very different proposition from the collective "want, desire, need, demand" syndrome. If we are wise enough to curb our appetites and to embrace the concept of necessity instead of want, the Earth still has, I believe, enough resources to nurture us while we relearn how to nurture the Earth. Nurturing the Earth brings us to the concept of reversibility.

Are the Consequences of Our Decisions Reversible?

Although sustainability potentially requires absolute reversibility in our decisions and subsequent actions if they prove ecologically and/or socially nonsustainable, the consequences of our decisions are revers-ible only in varying degrees in space and time. To understand this, let's consider three examples: a ditch, a dam, and soil.

A Ditch

If one wants to drain a swamp, one can dig a series of ditches, which lead water away from the swamp, and effectively lower the level of the water table. In so doing, one changes the habitat from a swamp to a meadow, a field, or something else, but only as long as the ditches remain functional.

If, however, one wants to reclaim the swamp, one must refill the ditches. While the water table will rise accordingly in the short term, it will take much longer for the life of the swamp to return. And because the swamp habitat has been set back to its beginning, it will never be as it would have been had the swamp not been drained. For all intents and purposes, however, this is an example of reversibility.

A Dam

Although dams can afford considerable economic and social benefits, their placement and construction must be grounded in sufficient knowl-edge of the river and its catchment basin to account for long-term ecological consequences. Dams are highly individualistic, and the same physical circumstance may elicit dramatically different responses. The effects of a dam in time and space can be considerable and may become apparent only after a long time.

While I was working as a vertebrate zoologist with a scientific expedition in Egypt in 1963 and 1964, a representative of the Egyptian Ministry of Agriculture spent time with us as we worked just north of the Sudanese border along the Nile. One day three of us from the expedition tried to help this man understand that building the High Aswan Dam across the Nile River was an ecological mistake. He could not, however, see beyond the storage of water for the generation of electricity, which was the official position of the government.

We explained to the government representative that building the dam would increase the geographical distribution of the snails that carry the debilitating disease schistosomiasis, a tiny blood fluke, from below the Aswan Dam (built by the British in the early 1930s) south to at least Khartoum in the Sudan, several hundred miles above the new, yet to be completed dam. At that time, it was still safe to swim above the Aswan Dam, where the water was too swift and too cold for the snails to live, but it was not safe to swim, or even catch frogs, in the water below the dam, where the snails already lived.

We told him that the Nile above the high dam would fill with silt, which would starve the Nile Delta of its annual supply of nutrient-rich sediment and affect farming in a deleterious way. We also told him that the dam could easily become a military target for the Israelis, as German dams were targets for the British during World War II.

The engineers building the new Aswan High Dam had intended only to store more water and to produce electricity, which they did. However, deprived of the nutrient-rich silt of the Nile's annual flood waters, the population of sardines off the coast of the Nile Delta in the Mediterranean diminished by 97 percent. In addition, the rich delta, which had been growing in size for thousands of years, is now being rapidly eroded by the Mediterranean, because the Nile is no longer depositing silt at its mouth.[43]

Until the Aswan High Dam was built, the annual sediment-laden waters of the Nile added a millimeter (a little less than a sixteenth of an inch) of nutrient-rich silt to the farms along the river each year. Now that the floods have been stopped by the new dam, the silt not only is collecting upriver from the dam, thus diminishing its water-holding capacity, but also is no longer being deposited on the riverside farms, thus decreasing their fertility. Soon the farmers will have to buy commercial fertilizer, something most of them probably cannot afford. In addition, because irrigation without flooding causes the soil to become saline, the Nile Valley, which has been farmed continuously for 5,000 years, will within a few centuries have to be abandoned. Also, schistosomiasis has indeed spread southward to the Sudan.

And a people, the Nubians, were displaced and their culture thereby destroyed. The Nubians whom I got to know were tall, straight, slender, very black with fine features, and lived many miles south of Aswan on small farms sandwiched between the east bank of the Nile and the Eastern Desert. Their village was neat and clean, and each house was decorated by embedding plates, with designs around their borders, into the outer mud coating of the doorways. The people had a wonderful sense of humor, were quick to laugh, and seemed genuinely pleased that I delighted in playing with their children and vice versa. For their part, the children seemed to have a good sense of self and of each day as an adventure to be lived to the fullest.

But the Aswan High Dam changed all that. The Nubians were moved inland from the bank of the Nile, whose quiet flowing waters and silent guardian desert had been a part of their lives seemingly forever. In place of their freely spaced, cool, airy, self-designed, and self-constructed homes, they were put into government-built, look-alike, minimum-quality housing, where many of them could not adjust and simply died.

There is yet another consequence of the Aswan High Dam, one I would never have thought of, even though I had studied the mammals along the Nile. The Nile annually flooded the many nooks, crannies, and caves along its edge, killing the rats whose fleas carry bubonic plague. Because the floods no longer occur, the rat population has soared, and bubonic plague is once again a potential threat.

I learned about this unexpected consequence of the Aswan High Dam from Dr. Wulf Killmann of the Deutsche Gesellschaft für Technische Zusammerarbeit, whom I met in Malaysia. As we visited about the effects of dams on rivers and oceans, I told him about my experience in Egypt. Dr. Killmann then told me that he had been part of a project to figure out how to control the ever-growing population of rats, which has become a serious health problem.

What would happen if the Aswan High Dam was removed? While the floods would once again begin fulfilling their many ecological roles, there would be an immediate problem of how to deal with all the silt trapped behind the dam. Then there is the question of what to do with all the steel, concrete, and other materials of which the dam is built. What would happen to all the economic investments and technological developments that have over the years sprung into existence because of the dam? How would the Egyptian people replace the social benefits engendered by the dam? And there is at least one benefit: greater electrical capacity for the country.

Even if all conceivable questions could be answered and most of the effects could be to some extent reversed, there is at least one that is final. The Nubian culture, in which I found such beauty and joy, would still be extinct. So the question is: How reversible in reality are the effects of the Aswan High Dam?

Let's consider another dam, the Bakun Dam, in the Malaysian state of Sarawak on the island of Borneo, which is slated for completion by the year 2005.[44] The plan for the Bakun Dam in the upper basin of the Sungai (=river) Rejang was shelved in 1990 but revived in 1994.

Upon completion, the area of the dam's water catchment will be 3.75 million acres. The reservoir will cover 173,750 acres of forest. Although much of the forest has already been degraded by logging and shifting cultivation, the rest is relatively pristine and rich in wildlife. The reservoir will also displace the aboriginal peoples of the Kenyah, Kayan, Lahanan, Ukit, and Penan tribes from their home lands, which are scattered in the upper area of the Sungai Balui, a tributary of the Sungai Rejang on which the dam will be located.

How reversible will the effects of the dam be on the habitat, the wildlife, or the people? Clearly, the habitat will be flooded and much of the remaining wildlife will be drowned. But what about the people? Where will they go? What will happen to their sense of place, their cultures, their identities as distinct peoples? Who will move them and how? Will they really understand why they are being moved? What kind of area will they have to live in? Will they ever regain their ancestral connection with the land if they are forced to move to an alien piece of ground? Do they have either a voice or a choice in the matter? If not, why not? Who makes these decisions? Many more questions could be asked in addition to these.

Finally, the great irony is that most of the electricity to be generated is for export. But what if the countries for which the electricity is intended build dams of their own and neither need nor want the exported electricity? How does one tell a whole people, a whole culture, that their displacement was an error, that they have been stripped of their ancestral home, their roots, for nothing? To what degree are the effects of this dam reversible? (For further information on dams in rivers and their effects, see "Forging a Sustainable Water Strategy."[45])

Soil

Because economists in our linear capitalistic system refuse to accept intrinsic ecological value as "real" value, we guide the use and attempted management of our natural resources only by the cost–benefit analysis of their potential economic value when converted to something else, such as from trees to boards, forage to red meat, water to hydroelectric power, and so on. This means that the only value economists can see is short-term specialization, a view that kills the very soils on which we depend for sustenance.

Many cultures have emphasized the trusteeship of the soil through religion and philosophy. Abraham, in his covenant with God, was instructed: "Defile not therefore the land which ye shall inhabit, wherein I dwell."[46] Chinese philosopher Confucius saw in the Earth's thin mantle the sustenance of all life and the minerals treasured by human society. And a century later, the Greek thinker Aristotle viewed the soil as the central mixing pot of air, fire, and water that formed all things.

In spite of the durability of such beliefs, most people cannot grasp them because they are intangible. The invisibility of the soil is founded in the notion that it is as common as air and therefore is taken for granted, as is air. To many people, the soil seems "invisible." But when we think about it, we realize that humanity, indeed human society, is somehow tied to the soil for reasons beyond measurable materialistic wealth.

Even though we can justify soil protection economically, our ultimate connection with it escapes many people. One problem is that traditional linear economics deals with short-term tangible commodities, such as fast-growing row crops, rather than with long-term intangible values, such as the future prosperity of our children. But when we recognize that land, labor, and capital are finite and that every system has a carrying capacity, one that depends on natural or artificial support, the traditional linear economic system becomes more like a cyclic biological system.

In the late eighteenth century, Thomas Malthus proposed that the human population would grow faster than the soil's ability to sustain it, but agronomic advances in this century led many shortsighted leaders to dismiss this idea as simplistic and overly pessimistic. Today, however, Malthusian theory seems prophetic when one considers the air pollution that poisons the soil, overgrazing by livestock and the growing desertification, global deforestation and the loss of the soil's protective cover of vegetation and its vitality, and the ensuing famines.

Those who analyze the soil by means of traditional linear economic analyses weigh the net worth of protecting the soil only in terms of the expected short-term revenues from future harvests, and they ignore the fact that it is the health of the soil that produces the yields. In short, they see the protection of the soil as a cost with no benefit because the standard method for computing "soil expectation values" and economically optimal crop rotations commonly assumes that the productivity of the soil will either remain constant or increase—but never decline.

In reality, however, reducing the productivity of the soil on marginal sites, the logic of which is both shortsighted and flawed, can push the expected present net worth of subsequent harvests below zero. Given that reasoning, it is not surprising that those who attempt to manage the land seldom see protection of the soil's productivity as cost effective. If we could predict the real effects of management practices on long-term economic yields, we might have a different view of the invisible costs associated with poor care of the soil.

One of the first steps along the road to protecting the fertility of the soil is to ask how various management practices affect the long-term productivity of the ecosystem, particularly that of the soil. Understanding the long-term effects of management practices in turn requires that we know something about what keeps the ecosystem stable and productive. With such knowledge, we can turn our often "misplaced genius," as soil scientist Dave Perry rightly calls it, to the task of maintaining the sustainability and resilience of the soil's fertility. Protecting the soil's fertility is buying an ecological insurance policy for our children.

After all, soil is a bank of elements and water that provides the matrix for the biological processes involved in the cycling of nutrients, which are elements under the right conditions of concentration and availability. In fact, of the sixteen chemical elements required by life, plants obtain all but two (carbon and oxygen) from the soil. The soil stores essential nutrients in undecomposed litter and in living tissues and recycles them from one reservoir to another at rates determined by a complex of biological processes and climatic factors. In a forest, for example, the losses of nutrients in undisturbed sites are small, but some are lost when timber is harvested. Others may be lost through techniques used to prepare the site for planting trees, reduce the hazard of fire, or control unwanted vegetation.

The resilience of forested sites following a disturbance, such as harvesting timber, is thus at least partly related to the ability of the soil to retain nutrients and water and to maintain its structural and biologically functional integrity during the period in which plants are becoming re-established. Beyond that, the health and fertility of the soil are reflected in the growth of the forest and the quality of the timber harvested now and in the future.

We would be wise, therefore, to reflect deeply on the observation of soil scientists V.G. Carter and T. Dale, who point out that civilized people despoiled their favorable environment mainly by depleting or destroying the natural resources; cutting down or burning most of the usable timber from the forested hillsides and valleys; overgrazing and denuding the grasslands that fed their livestock; killing most of the wildlife and much of the fish and other water life; permitting erosion to rob their farm land of its productive topsoil; allowing eroded soil to clog the streams and fill their reservoirs, irrigation canals, and harbors with silt; and in many cases, using or wasting most of the easily mined metals or other needed minerals. As a result, their civilization declined in the middle of its despoilation, or they moved to new land.[47]

With the above in mind, we need to pause and consider carefully the counsel in 1905 of the thinker George Santayana: "Those who cannot remember the past are condemned to repeat it." And if we are wise, we will ask: What are Nature's penalties for economically and ecologically disregarding the soil? One obvious penalty is loss of fertile topsoil.

In our concern for the topsoil, we need only take the lessons of history, because although the birth of agriculture caused civilizations to rise, it was abusive, linear agricultural practices that destroyed the topsoil and thus caused the collapse and extinction of civilizations. And yet, with all the glaring lessons of history spread before us around the world, with all our scientific knowledge, and with all our technological skills, we insist on walking the historical path of agricultural ruin and impending social collapse.

The supreme irony is that even as we work to rid ourselves of all nuclear weapons on Earth and to establish a lasting peace among humans, we continue to commit genocide by ruining our environment. Through rain and snow, airborne pollutants reach the entire Earth from the tops of the mountains that pierce the clouds, through the vegetation, down into the soil, and down to the deepest recesses of the sea. As we continue to poison our environment, we destroy the stage—the soil—on which the entire human drama depends for life. Destroy the stage, and the drama is no more.

Soil scientist J.C. Lowdermilk addressed this point when he wrote, "If the soil is destroyed, then our liberty of choice and action is gone, condemning this and future generations to needless privations and dangers." To rectify society's careless actions, Lowdermilk composed what has been called the "Eleventh Commandment," which demands our full and unified attention and our unconditional embrace if human society is to survive:

> Thou shalt inherit the Holy Earth as a faithful steward, conserving its resources and productivity from generation to generation. Thou shalt safeguard thy fields from soil erosion, thy living waters from drying up, thy forests from desolation, and protect thy hills from overgrazing by thy herds, that thy descendants may have abundance forever. If any shall fail in this stewardship of the land thy fruitful fields shall become sterile stony ground and wasting gullies, and thy descendants shall decrease and live in poverty or perish from off the face of the earth.[48]

History is replete with lessons, which point out that soil, once destroyed or lost, takes many human lifetimes to replace. Therefore, when sustainability is put in terms of human communities, any decision that either degrades the fertility of soil or causes the loss of soil must be considered irreversible.

All this discussion is to point out that sustainability is either conferred on human communities or denied human communities through a process called development. What does development mean? How does the process either give sustainability to or withhold it from a community?

DEVELOPMENT

Development, if it is sustainable, enables us to transcend our crisis in perception because it requires that decisions be based on assumptions different from those of the expansionist economic world view to which we have heretofore subscribed. In basing decisions on a unified systemic world view, society can guide itself toward behaviors and lifestyles that are environmentally sustainable and thereby ensure, to the greatest extent possible, its own cultural sustainability. This means, however, that development must be understood and accepted in terms other than the current linear notion of continuous material growth.

What Is Meant by Development?

Of the several facets reflected in the term "development," we in the United States have chosen to focus on a very narrow one: development as material growth through centralized industrialization, which we glibly equate with social "progress" and economic health. The narrowness of this view is, I believe, behind the notion of "developed" versus "developing" nations.

I have over the years worked in a number of countries without giving much thought to the notion of "developed" versus "developing" or, as some would put it, "underdeveloped," although I have spent time in both. During my recent trip to Malaysia, however, I was profoundly struck by the arrogance and the narrowness of such thinking.

Malaysia is the only place in which I have ever heard the people refer to their own country as "developing," as though they are somehow behind the "developed" countries and must somehow "catch up" to be equal. Yet the Malaysians have a national unity the likes of which I have never before seen, not even in the United States, where all my life I have been taught about and heard about an equality that is not practiced.

Malaysia is as great a mixture of cultures, national origins, and religions living in a small space as I have ever seen. Yet when I asked people what their ethnic background was, their answers—to a person—reflected national unity. They referred to themselves as Malaysian Chinese, Malaysian Indians, Malaysian Sri Lankans, and so on. Were I to ask such a question in the United States, however, the response would be Afro-American, Chinese-American, Japanese-American, German-American, Italian-American, and so on. While the difference may be subtle, it is profound. The Malaysians focus on their unity, while we in the United States focus on our sense of separation.

On any given day in Malaysia, I might eat my breakfast the Malay way, using both hands, with a spoon in one and a fork in the other. At lunch, I might eat with chopsticks, and at supper, I might eat as much of the world eats, with my right hand as the only utensil. There were even four hour-long evening news programs, one each in Malay, Indian, Chinese, and English. Of course there are social problems, but I have never before experienced such integration of differences into an amalgamative sense of wholeness.

As a guest and stranger in Malaysia, I felt that sense of wholeness encompass me. I felt welcomed and accepted for what I am—not who I am. In a strange, undefinable way, I felt more at ease and at one with the people of Malaysia as a whole than anywhere I have ever been.

If this is not development, I do not know what is! But then it depends on how one defines development. If development is defined as a certain material standard of living based on the consumerism of centralized industrialization, Malaysia is indeed behind the United States. But if development is defined as social civility and tolerance, the United States, compared to Malaysia (a nation even younger than the United States), is a *developing* country.

And what about aboriginal peoples who not only have civility and tolerance but also have a long-term sustainable relationship with their environment. Are they not developed?

It is ironic that the very people who consider themselves to be developed and therefore "civilized" are the ones who have so ruthlessly destroyed the cultures of those they unilaterally brand as "undeveloped" and therefore necessarily "uncivilized savages." Fortunately, despite the continuing onslaught of "civilized" peoples, there are a few remaining aboriginal ones, some of whom live in the deserts of Australia and the jungles of South America, as well as other parts of the world.

I say fortunately, albeit they are severely endangered, because there is much about development and sustainability that we in the industrialized world can *relearn* from them. After all, our ancestors were also indigenous tribal people at one time. Our problem of late is that we have forgotten most, if not all, of the wisdom they once knew. And it is precisely this loss of ancient wisdom that is forcing us to focus on a contemporary question.

How must we view development if the concept is to be equitable and sustainable? Make no mistake, "develop," "developed," and "development" must be viewed by all parties as equitable if development is ever to become sustainable.

Specialization Versus Generalization

Although the genus *Homo* emerged "only" five to eight million years ago, it has become remarkably adaptable and successful. Unlike most genera, which exist somewhere between five and ten million years before fading into extinction as other genera take over, we modern humans have no immediate threat of extinction to face. Unless, of course, we as a species kill ourself by making our home planet unfit for our own existence through pollution—something we seem well on our way to doing.

In this sense, whatever crisis and transformation await humanity in the future, it is not one of individuals but of communities and societies. We must therefore distinguish between crisis and transformation in a species and in a society.

The human being as a species can continue living on Planet Earth for thousands of years, but human society is sick and in crisis. The culmination of the crisis may well cause the extinction of today's dominant social systems (witness the collapse of the former Soviet Union) and the emergence of new ones. The point is that human beings are generalists while human societies are specialists. The distinction between the two is both precise and critical.

A generalist, in either a biological species or a nonbiological system, can survive under a wide range of environmental circumstances, can use numerous kinds of energy, and can either fit itself to a wide variety of conditions or fit a wide variety of conditions to itself. A specialist, on the other hand, is fitted to a highly specific set of circumstances within its environment and can derive and use only certain kinds of energy.

Let's look at coyotes versus wolves. Coyotes are exceedingly adaptable, independent animals that can seemingly survive anywhere, including the suburbs of Los Angeles, California. Coyotes roam the country singly, in pairs, and for part of each year as family groups. Ranging around alone, in pairs, or as family groups allows them to prey on a wide array of kinds and sizes of animals, beginning with grasshoppers (as pups) to prey as large as adult mule deer and yearling domestic cattle (as adults). Coyotes are also adept at eating fruits, and in some parts of their geographical distribution they are called "melon wolves" because they steal from the farmers' fields.

As a generalist, the coyote can survive under a wide range of environmental conditions, from Texas to Alaska and from the Pacific Northwest to the eastern seaboard. Their arrival in Alaska and on the eastern seaboard within recent years is due primarily to the clearcutting of vast areas of dense forest. Our social activities have opened up thousands of square miles for the coyote to inhabit, areas that it can readily take advantage of because of its extraordinary adaptability. And because of its wide array of food items, the coyote can make use of a goodly variety of sources of energy.

Wolves, on the other hand, are social animals that live in packs. Compared with coyotes, their group life places limits on their ability to use habitats. Thus, a far greater number of coyotes than wolves can live in Yellowstone National Park, because a pack of wolves acts as a single large organism and therefore requires a much vaster area in which to hunt. A pair of coyotes can live on rabbits and fruits in season, but a pack of five to seven wolves, each of which is much bigger than even a big coyote, acts collectively as a single large animal and in order to survive therefore requires prey the size of mule deer, elk, and moose.

A pack of wolves has a much harder time staying fed than does a pair of coyotes. It takes far more time, energy, and trials for a pack of wolves to select, chase, and bring down large prey at any time of the year than it does for a coyote, which at certain times of the year can do quite nicely on a diet of grasshoppers, meadow voles, and ripe berries.

As a specialist, the wolf is fitted to a narrow set of environmental circumstances and can survive only it if finds prey large enough to feed the pack as a whole. The wolf, therefore, has a limited range of prey items to which it is effectively adapted as sources of energy, and it can neither fit itself to a wide variety of conditions nor fit a wide variety of conditions to itself.

The wolf, as a highly adapted specialist, is vulnerable to extinction by societal desire, while the coyote, as a supremely adaptable generalist, is likely to outsurvive humanity itself. This is particularly evident as the wolf's geographical range shrinks in the face of societal pressures on the landscape and the coyote's geographical range increases in concert with those same pressures. How does this concept apply to humanity and human society?

We as members of the human species are about the most successfully adaptable generalists on Earth. People live in the frozen tundra and along the sea ice above the arctic circle, throughout the temperate forests and plains, in the hot deserts, and in the depths of steaming tropical jungles. We live and reproduce on every continent and at every latitude between the two polar circles and beyond, and we have found cures for enough diseases to vastly increase our numbers and our longevity. In addition, we are generalists in the social sense. We have built and live in societies ranging from nomadic food-gathering tribes to sophisticated post-industrial civilizations and from raw military dictatorships to "grass roots" democracies.

Although we as individuals and as a species, like the coyote, are adaptable generalists, our modern societies are becoming more and more rigid specialists, like the wolf, as we allow ourselves to become professionally specialized. I have observed, for instance, that a politician who "creates" jobs and takes credit for his or her accomplishment seldom accepts equal responsibility for the built-in obsolescence of those same jobs. Yet the more specialized the jobs are, the more certain is their built-in obsolescence. Ultimately, therefore, those people will again be out of work when the economy changes, when the need for their jobs disappears, or when the resources on which their jobs depend, such as old-growth forests and large stocks of fish, run out.

Because change is inevitable and because there are few "guaranteed" jobs for specialists, such as undertakers, over the long term, one is wise to remain adaptable. Nevertheless, specialists appear to have an advantage in the job market in the short run, while generalists have a corresponding disadvantage. But generalists have a great advantage in the long run because rather than being rigidly set in their ways (as are most specialists), they are versatile and open to learning many things.

Such versatility allows generalists to flow with unforeseen changes and makes them rich in the experience of life, while specialists become encrusted in their specialties and are increasingly cut off from the experiences of life.

To the extent that individuals become rigid specialists, communities and society at large become rigidly specialized. This specialization is increasingly apparent through the unfolding impact of the global division of labor and the polarization of global politics. In contrast, a Stone Age village and its neighbors could care for every basic need of its people. Short of such catastrophic disturbances as floods and volcanic eruptions, the Stone Age folk could cope with the vicissitudes of Nature.

Today, however, fewer than perhaps a dozen societies can produce enough food to supply the necessities of their own population, and the same can be said of energy, water, wood products, transportation, and communication, not to mention the myriad consumer goods most people seem to regard as their birthright. When societies are economically specialized and interdependent, they are more or less at the mercy of other societies for the items they lack. Japan, for example, offers high technology in exchange for almost everything else it needs to survive as a modern society. An example of the other side of the coin is Gambia in West Africa, whose economy depends on the export of groundnuts. There, a crop failure spells economic disaster.

The interdependence of societies in the political arena is much more obvious. No country today, including the United States, thinks it can any longer assure its own materialistic defense without military allies, strategic bases, earth-circling spy satellites, and networks of collaborating intelligence agencies—all of which create and maintain what is increasingly apparent to be not peace but a balancing act of global terror.

The increasing specialization in almost all dimensions of contemporary societies, like that of a pack of wolves, makes them more and more vulnerable to collapse from economically powerful competition and/or sudden changes in the environment itself. Specialists are unstable and subject to extinction (unemployment) as their usefulness comes and goes, whereas generalists continually improvise, adapt, and adjust to new ways.

Modern human societies are lured into specialization by the linear thinking of quick monetary gains and materialistic security implied by industrialization. Today's societies, each strongly interdependent on one another as they live side by side, have evolved into an inherently precarious social system. When we add to this already precarious system the introduction of new technologies of production and communication, the system becomes risky in the extreme, and the imminence of a system-wide social crisis should come as no surprise.

Since World War II, social specialization has been global. Society has become specialized and interdependent in the way it extracts resources from Nature; in the way it cultivates Nature; in the way it uses energy, food, and raw materials; in the way it builds dwellings and cities; and in the way it disposes of its own wastes. This is an unstable social system at risk of extinction; again, witness the collapse of the former Soviet Union. Another system could take its place, however, because humanity itself is a generalist and is not condemned to live and die in the super-specialized societies it has created in the postwar era.

In the words of White Eagle: "Whatever man thinks he becomes. What he sees in his surroundings, in his work, in his religion; whatever it is he creates, he is in it....Man is his own jailer, his own liberator." And in the words of James Allen: "A man is literally *what he thinks*, his character being the complete sum of all his thoughts."

Sustainable Development

The problems in our communities can neither be isolated nor understood without first understanding their context. This means that one must understand the various parts of a community and their interactions before one can understand why something is the way it is.

In developing this understanding, process has primacy over the parts because process directs the outcome, the function of each part. A part does not control the process, although it may influence it.

In placing development within a new context of conscious choice, the answers to such questions as "what do we mean by development," "what is underdeveloped," and "what is poverty" will be very different than they are today. If a lifestyle promotes sustainability through conscious choice, conscious simplicity, and self-provisioning and recognizes the relationships between one's own sustenance and the livelihood of one's immediate surrounding (one's fidelity to one's sense of place) in relationship to the larger world, that life is not necessarily perceived as one of poverty. This leaves the way open to change the indicators of development.

Progress, therefore, would be any action that moves a person, community, culture, or society toward social/environmental sustainability. For society to progress, decisions must be made that recognize and respect the requirements and rights of future generations, as well as the requirements and intrinsic value of all species and the Earth's carrying capacity with respect to its human population. (Carrying capacity is the number of individuals that can live in and use a particular landscape without impairing its ability to function in an ecologically specific way.) This position is very different from our blind faith in material progress, which we think of as development.

Again, I think the narrowness with which we view development (i.e., the centralized production of material consumer goods through industrialization) is one root of the arrogance with which Western industrialized countries designate themselves as "first world" nations and all the others as "second" or "third world" nations. In Canada, however, the aboriginal peoples have turned this notion around.

Although the Canadian government refers to the aboriginal peoples as "Indian bands," the people think and speak of themselves as "First Nations." The people think of themselves as First Nations because they were the original people, the indigenous people on the land in time and space long before any of the outside invaders even knew that the "New World" existed.

Add development to this time/space sense of First Nation, and a clearer picture emerges. The indigenous peoples were not only the first humans in what is now Canada but also had developed a lifestyle that had long been sustainable in and with their environment, despite the fact that they warred amongst one another. Yes, the invaders— with greater numbers and more destructive technology —subdued the indigenous peoples, stole their land, and systematically destroyed their cultures. But these same invaders, upon landing on foreign shores, began immediately destroying the environment through economic exploitation for personal gain, something the aboriginal peoples were not prone to do. In fact, the invaders even fought wars amongst themselves over who was going to get which of the stolen spoils.

There is a great contradiction here in the notion of development. Those invading peoples who deemed themselves more advanced or more developed than the indigenous peoples destroyed lifestyles that had been, more often than not, sustainable for millennia, while simultaneously introducing lifestyles of exploitation for personal economic gain that have proven to be nonsustainable. If, therefore, social/environmental sustainability is added as a necessary component to the concept of development in the broader sense, the indigenous Canadians have an even greater claim to being the "First Nations," and so do all other indigenous peoples in the world.

But who, then, are the "second" and "third world" countries? Professor Ralph Metzner of the California Institute of Integral Studies has a good idea.[49] He suggests that the world of modern cities and the nation state is the second world country, while the global capitalist/communist industrial economy constitutes the third world country.

Historically, he says, each of these worlds *superimposed* itself on earlier cultures (in the sense of absolute force), and in an ecological sense, these later, larger systems became parasites, which destroyed the indigenous cultures they parasitized. This parasitism was—and is—largely in terms of the flow of energy. "The flow of resources, including raw materials and food," observes Metzner, "is primarily from the indigenous world to the urban, national, and...industrial worlds, whereas military and political control is exerted in the opposite direction."

Sustainable development is thus about the notions of *enoughness* and *reversibility*. Here the operative questions are: When is enough enough? and If we err in our decision, is the outcome reversible? Such questions are crucial because sustainable development is necessary to promote a change in the content of social/environmental decisions. What is needed to resolve our social/environmental problems goes beyond environmentally safe commodity production and technology.

Instead of the current tinkering with symptoms of our social/environmental malaise, problems must be solved at their source—world view assumptions and values—because these drive our decisions, policies, and plans. Sustainable development therefore questions the very purpose of society and our participation with our home planet, and demands social/environmental justice, which challenges the very heart of our perceived relationship with Nature and one another, present and future.

We as planetary citizens must learn to think at least seven generations ahead when making decisions, because the great and only gift we have to give those who follow is *potential choices and some things of value from which to choose.* Today's decisions become tomorrow's consequences, a notion that highlights the word "responsibility."

Responsibility is a double-edged sword in that our responsibility, our moral obligation, is to choose carefully today so that the generations to come can respond viably to the circumstances we have created for their time of choice. Intelligent decisions on our part are possible only when we both recognize and accept the intrinsic value of Nature as a living organism rather than accepting Nature only as a collective resource (host) from whose body we extract (parasitize) a variety of commodities as the life's blood of our dysfunctional, linear economic system.

Development must be flexible and open to community definition because the values promoted must meet various needs and situations in space and time while safeguarding sustainability. The process of valuation embodied in sustainable development addresses social/environmental justice in recognizing the necessity of equal access to resources as well as equal distribution of goods and services while simultaneously protecting the long-term ecological sustainability of the system that produces them.

Sustainable development also addresses the need to promote education and feelings of self-worth in people, allowing them to act as catalysts in the process of change, whether in their own lives or in the life of society. For change to be a creative process, each person must respect every other person as well as the intrinsic value of his or her environment.

Finally, the valuation/decision process through which sustainable development works must flow within and promote the democratic frame of reference because democracy only works when it is actually practiced. In this sense, most of the change must be directed by the people from the bottom up—the "grass roots" of the local community.

At What Scale Is Sustainable Development Possible?

Although we, as a society, do not have many answers, we can find guiding principles for action in the questions we ask. For example: At what scale is sustainable development possible? Thinking about the scale at which sustainable development is possible calls to mind a catchphrase: think globally; act locally. But when considering the notion of global community from an individual's standpoint, helping to heal the world seems like a hopeless task, even an unintelligible abstraction. Yet as people come to understand their effect on the Earth, an increasing number of them want to do something to change the worn-out expansionist economic paradigm. But if global and national strategies are abstractions too far removed from the average person's realm of experience, at what level can the average person act and be effective?

Sustainable development must be implemented where people are invested in a sense of place and are able to learn, feel, and be empowered to act—the local level. Sustainable development must be integrated into policies and decisions in local communities where people have the power to effect change and make decisions based on a unified systemic world view, a "first world" view, one that begins healing the environment in the present for the future.

Here, as he so often does, author Wendell Berry cuts to the core: "That will-o'-the-wisp of the large-scale solution to the large-scale problem, so dear to governments and universities and corporations, serves mostly to distract people from the small, private problems that they may in fact have the power to solve. The problems, if we describe them accurately, are all private and small. Or they are so initially."[50] It is thus imperative that we address the fundamental causes of our problems at their roots—our thinking and behavior at the level of the local community—or we will always be dealing with symptoms and band-aid solutions that compound the problem by denying the cure.

Through the concept and practice of local community, people can empower themselves and support one another through decisions that promote a sustainable world as well as increase their quality of life. Within a local community, people can act as the force driving change in their political system even as they alter their lifestyles. Through their actions, people partake in guiding destiny, despite the fact that a local community is part of a larger more impersonal governmental system.

Local community governments, as opposed to county, state, or national governments, have both a greater degree of understanding and the interaction necessary to amend local problems, such as land use, waste reduction, political representation, and education. They are therefore better able to implement and adjust to aspects of sustainability in the social/environmental arena.

People both define their local communities and are defined by them in that communities play a primary role in maintaining cultural values within and among generations. The collective of individual values determines familial values; the collective of familial values determines community values and determines what is appropriate behavior, poverty, and success. As we grow up and are taught at home, educated in schools, and participate in community, socialization occurs, norms are set, and societal control takes place.

People make most of their decisions, do most of their consuming and waste production, and develop many personal relationships within their local community. It is not surprising, therefore, that lifestyle becomes a political issue.

Citizens at the local level can also begin drawing connections between personal consumption and its effects on local, regional, national, and global economic well-being and environmental health. Although political pressure must be exerted continuously on national governments, both at home and abroad, lifestyles in each and every local community have direct and immediate effects on the biosphere.[51]

Local communities, through their collective effects on world society, act as catalysts for change in society at the local, regional, state, and national levels and finally in the world itself. Through the behavior of their individuals, local communities contribute greatly to environmental health and the global climate. And because we as individuals collectively comprise local communities, which in the collective comprise society at the regional, state, and national levels, we can heal our global environment simply by changing our individual behaviors. Local communities, as the force that drives change for better or worse, are thus the appropriate scale for dealing with sustainable development.

SUSTAINABLE COMMUNITY DEVELOPMENT

C
ommunity, in the context of sustainability, is a group of people with similar interests living under and exerting some influence over the same government in a shared locality. They have a common attachment to their place of residence, where they have some degree of local autonomy. People in a community share social interactions with one another and organizations beyond government and through such participation are able to satisfy the full range of their daily requirements within the local area. The community also interacts with the larger society, both in creating change and in reacting to it. Finally, the community as a whole interacts with the local environment, molding the landscape within which it rests and is in turn molded by it.

The components of community will change when the social focus shifts from the expansionist economic world view to a unified systemic one, as still exists in the few remaining unspoiled aboriginal cultures. When the world view becomes unified and systemic, people will realize that they are but one species among the many in a dynamic, ever-changing, interactive, interconnected, interdependent system, and a new sense of community will encompass all living things, including soils, within the common area.

Historically, however, our vision has been directed by competition. As yet, we are still so overdependent on and mesmerized by competition that it is our predominant model for learning and change. There is nothing intrinsically wrong with competition; it can even be fun and promote invention and daring. Our problem is that we have lost the balance among competition, cooperation, and coordination at precisely the time we most need to work with one another. We thus find ourselves oftentimes competing with the very people with whom we need to collaborate.[40]

In a unified systemic world view, a local community serves five purposes:[52] (1) social participation—where and how people are able to interact with one another to create the relationships necessary for a feeling of value and self-worth; (2) mutual aid—services and support offered in times of individual or familial need; (3) economic production, distribution, and consumption—jobs, import and export of products, as well as the availability of such commodities as food and clothing in the local area; (4) socialization—educating people about cultural values and acceptable norms; and (5) social control—the means for maintaining those cultural values and acceptable norms.

With the current disintegration of family and community in American life, it is unlikely that most people in this country really have an intimate sense of belonging. We have largely lost our sense of connection to and with community. One reason for this lack of community may be our lopsided expansionist economic world view in which material possessions take the place of spirituality, as manifested in quality relationships and mutual caring. If, however, human society and its environment are ever to become sustainable, it is necessary to rediscover or recreate our sense of local community in order to balance the material with the spiritual, the piece with the whole, which is the essential balance required in a unified systemic world view.

Although the last two centuries may have nurtured such institutions as freedom and equality, they have done little for the fraternity and solidarity that hold societies together. But this softer value is the social capital that enables people to work together, to trust one another, to commit to common causes, and is absolutely critical to the success of community in the fulfillment of its vision.[53]

For a community to fulfill it vision, it must be grounded in personal ethics, which are translated into the social ethics of a community. This puts the responsibility for one's own conscience and behavior where it rightfully belongs—squarely on one's own shoulders. With a strong sense of personal and social ethics, communities will be spared wasting time and money on policing socially unacceptable behavior. With a strong sense of personal and social ethics, neither the environment nor future generations will be the dumping ground for personal and social irresponsibility.[53]

While for some, community may simply be a useful new concept to wrap around old ideas and institutions, for others it will be a new set of ideas, a new frame of reference about how people relate to one another and to the wider world. Its value lies in making a bridge between people's core values and principles for action and governance, which will help shift perceptions about what politics and government are really for. Community also changes one's thinking about the scale of public action because the scale of effective organization has shrunk to that of the school, neighborhood, and interest group.[53]

Community is a way of valuing the independent voluntary or nonprofit organization. Community rests on such intermediate-level institutions as neighborhood schools, family centers, or volunteer organizations for its expression, things that the top-down models of local government cannot fulfill.[53]

There is an increasingly common yearning for more defined ethical values with which to fill politics with meaning and purpose. The language of community is one way of doing this, of reconnecting people with a set of shared values and principles with which to embrace the uncertainties of life. Ethics must therefore be nurtured as one of the most valuable assets in making human communities work.[53]

LOCAL COMMUNITY DEVELOPMENT

Community is a deliberately different word than society. Although it may refer to neighborhoods or workplaces, to be meaningful it must imply membership in a human-scale collective. It must be a place where people encounter one another face to face. Community must therefore nurture human-scale structural systems within which people can feel safe and at home.[52]

Local community development is a process of organization, facilitation, and action that allows people to create a community in which they *want* to live through a conscious process of self-determination. If that which they create is in social/environmental harmony, then their legacy is one of sustainable opportunities for their children. Community development is thus a process in which the ideals of sustainable development can be implemented by both allowing and encouraging people to act as catalysts for sustainable social change at the community level.

Local community development, which is the democratic process at the local level of our domestic lives, may well be the instrument through which people can create the means to invigorate society. Given the current state of our economy, politics, disintegrating families, violent social relations, uncertain sense of security, and the confused condition of our guiding values, it may be that community development is the best opportunity for applying the democratic process.[52]

Community development is the mechanism through which people empower themselves by increasing their ability to control their own lives in order to create a more fulfilling existence through mutual efforts to resolve shared problems. Community development works based on the belief that through collective action people can successfully resolve their issues as well as organize and implement change. It thus *promotes a sense of accomplishment and belonging through shared learning and service.*

Another way local community development enhances people's potential is by helping them dissolve barriers. Barriers can be dissolved by bringing all parties affected into the decision-making process. While prejudice and a sense of inequality suppress relationships among people due to their perceived differences, community development helps them learn to cope collectively with the many problems that affect their lives by uniting them in common cause when they might not otherwise choose to associate with one another.

Because local community development is a democratic process that works only when it is accessible to and implemented by the majority of the population, it is necessary to involve as many members of a community as possible in the process of improving democracy through participation. The more diverse the participants are in the democratic process of community development, the more accurately the community will be represented, the greater will be the sense of equality in rights and duties, and the truer the outcome. Local people are thus empowered by acting collectively through organizations to influence decisions, policies, programs, and projects that affect them as a community.

The important word here is "empowerment," which addresses people's perceived capacity to influence decisions that affect their lives through active participation in and hence improvement of the democratic process. Central to the notion of empowerment is people's willingness to accept responsibility for their own behavior, first by overcoming interpersonal barriers and learning to work with one another for a greater good and second by directing formal authority within the democratic process.

LOCAL COMMUNITY AND DEMOCRACY

Democracy is the backbone of local community development and ultimately community sustainability. It is thus critical to understand something about democracy as a practical concept. Democracy is a system of shared power with checks and balances, a system in which individuals can affect the outcome of political decisions. Democracy is designed to protect individual freedoms within socially acceptable relationships with other people individually and collectively. People practice democracy by managing social processes themselves. Democracy is another word for self-directed social evolution.

Democracy in the United States is built on the concept of inner truth, which in practice is a tenuous balance between spirituality and materialism. One such truth is the notion of human equality, in which all people are pledged to defend the rights of each person, and each person is pledged to defend the rights of all people. In practice, however, the whole endeavors to protect the rights of the individuals, while the individuals are pledged to obey the *will* of the majority, which may or may not be just to each person.

The "will" of the majority brings up the notion of freedom in democracy. Nothing that I know of in the Universe is totally free; rather, everything expresses freedom within some sort of limits. Just as there is no such thing as a truly "free market" or an "independent ecological variable," so are individual and social autonomy protected by moral limits on the freedom within which individuals and society can act. To this effect, author Anna Lemkow[54] lists four propositions of freedom: "(1) An individual must win freedom of will by self-effort, (2) freedom is inseparable from necessity or inner order, (3) freedom always involves a sense of unity with others beyond differences, (4) freedom is inseparable from truth—or, put the other way around, truth serves to make us free."

Lemkow goes on to say:

> We tend to think of freedom as dependent on circumstantial or external factors, but these propositions point us inward, suggesting rather that freedom is a state of consciousness and…depends on ourselves. Indeed it is something to be won, something to be attained commensurately with becoming more truthful, or more attuned to and aligned with the abiding inner, metaphysical, or moral order or law.
>
> Socio-political and economic freedom or liberty, in turn, would depend (at least in the longer term) on the predominant level of consciousness of the citizenry.

Lemkow is positing that a human being is not completely free to begin with, but possesses the potential capability of self-transformation in the direction of fuller freedom. Beyond this, democracy requires respect for others and excitement in the exchange of ideas. People must learn to listen to one another's ideas, not as points of debate but as different and valid experiences in a collective reality. While they must learn to agree to disagree at times, they must also learn to accept that, like blind people feeling the different parts of an elephant, each person is initially limited by his or her own perspective. When these things happen, people are engaged in the most fundamental aspects of democracy and come to conclusions and make decisions through participative talking, listening, understanding, compromising, agreeing, and *keeping* their agreements in an honorable way.

In a democracy, *connection* and *sharing* are central to its viability because a democracy only works when it is being practiced. And a democracy can only be practiced by an educated public. A sound education is therefore an absolute necessity for the survival of democracy. Withhold education and dictatorship is a virtual certainty.

In a democracy, people are not required to separate feelings from thoughts concerning a topic. Their roles—as teachers, students, leaders, facilitators, and followers—fluctuate within and across the issues. The importance of a democratic system lies in its connection to people's lives, their own experiences, and the real problems and issues they daily face. Practicing democracy can be thought of as education in and for life.

The challenge, therefore, is to engage people in the democratic process, which is difficult when they confuse the government with the administration of the government or when the administration becomes so dysfunctional that people despair in their seeming inability to fix it. To this end, Thomas Jefferson made the following observations concerning the federal government of the United States:

> When all government, domestic and foreign, in little as in great things, shall be drawn to Washington as the center of all power, it will render powerless the check provided of one government on another, and will become as venal and oppressive as the government from which we separated [1821].[55]
>
> If ever this vast country is brought under a single government, it will be one of the most extensive corruption, indifferent and incapable of a wholesome care over so wide a spread of surface. This will not be borne, and you will have to choose between reformation and revolution. If I know the spirit of this country, then one or the other is inevitable. Before the canker is become inveterate, before its venom has reached so much of the body politic as to get beyond control, remedy should be applied [1822].[55]

Although I know of no perfect government, a just government must be founded on truth, not knowledge—something we in the United States have all too swiftly forgotten. To achieve such government, it must be based on service (where people are other-serving) rather than power (where people are self-serving). Therefore, development is only sustainable and environmental protection is only possible if the government is accountable to its people beyond special interest groups and political lobbyists.

We live in an increasingly complex society of intense competition and materialism. In such a society, people commit evil acts in order to gain power, both through positions of authority and financial success. People commit evil acts while falsely expecting to benefit by them, thinking that such benefits will somehow bring happiness. But in the end, as Socrates warned, the guilt of the soul outweighs the supposed material gains. Thus, because people lack perfect knowledge and perfect motives, democracy must be continually practiced and continually improved through that practice.

Nevertheless, the *people* are the government, but they can govern only as long as they elect to use the constitutional system for empowerment at the local community level. It is important to understand that empowerment is personal self-motivation. No one can empower anyone else; one can only empower oneself. One can, however, give others the psychological space, permission, and skills necessary to empower themselves and then support their empowerment. Beyond that, one can help in the process of empowerment and can increase the chances of success by recognizing another's accomplishments each step of the way. That is true democracy.

Therefore, in the case of the United States, where the government is of the people, by the people, and for the people, when the people empower themselves, they are the government, and it is the administration of that government and not the government itself—that resides in Washington, D.C. The administration becomes the government when the people turn their power over to the administration and in effect say: "I am a victim and cannot change the system," or "take care of me." When this happens, democracy is endangered, as noted by Professor Alexander Tayler over 200 years ago, when what is now the United States was still a British colony. Tayler penned the following:

> A democracy cannot exist as a permanent form of government. It can only exist until voters discover that they can vote themselves largesse from the public treasury. From that moment on, the majority always votes for the candidates promising the most benefits from the treasury, with the results that a democracy always collapses over loose fiscal policy, always followed by a dictatorship.
>
> The average age of the world's greatest civilizations has been two hundred years. These nations have progressed through this sequence:
> from bondage to spiritual faith;
> from spiritual faith to great courage;
> from great courage to liberty;
> from liberty to abundance;
> from abundance to selfishness;
> from selfishness to complacency;
> from complacency to apathy;
> from apathy to dependence;
> from dependence back again into bondage.[56]

If what Tayler says is true, then I think we are on the down-side of the democratic cycle, a thought echoed in 1993 by William Bennett, former secretary of education: "It's a misdiagnosis to say [America's] problem is economics. It's a cultural decline. Our problems are moral, spiritual, philosophical, and behavioral."

But that does not mean that the cycle is irreversible. I think the cycle can be altered for the better by conscious choice. Democracy can be a viable system by inviting a constant reinterpretation of itself based on asking morally, socially, environmentally, and economically sound, farsighted questions. In this way, it is always in a state of becoming, which continually interweaves it within the intimacy of life.

For democracy to remain viable, its principles and processes must be used, because they form an interconnected, interactive system of balancing and integrating contrasting perceptions of data, fact, and truth. A working democracy is thus predicated on finding the point of balance through compromise in such a way that the rifts between opposites can be minimized and healed, including those that deal with the various scales of economy.

LOCAL COMMUNITY AND ECONOMY

Community development and economic development differ in that economic development is only one aspect of community development just as economy is only one aspect of community. Economic development works to increase activity and stability in the production, distribution, and consumption of products and services within a community, but that is not enough. According to economists Robert Constanza and Herman E. Daly: "To effect a true synthesis of economics and ecology [and hence community] is the second most important task of our generation, next to avoiding nuclear war. Without such an integration we will gradually despoil the capacity of the earth to support life. Gradual despoilation is certainly preferable to destroying it all at once in a nuclear war, but is still an unhappy prospect."[40]

Thus, while it is important to recognize that economic development contributes to community stability, its primary focus is to increase economic activity within the local area through business retention, expansion, and attraction, as well as job training. It must of course be recognized that economic development within the context of sustainable community development must be in harmony both with the productive capacity and integrity of the environment over time and with human dignity and a sense of well-being. Community development, on the other hand, focuses on increasing socialization, mutual aid, economic activity, and social participation and control, which in turn increases the social, educational, and cultural stability of a community.

Community development—planning the local process (vision, goals, and objectives are discussed in *Resolving Environmental Conflict*[25])—is a necessary step before an economic development plan can be successfully implemented. This becomes clear when a community tries to identify its vision and goals for a business plan, which must be related to and interact with all aspects of community development. Many skills are necessary to organize people, run meetings, and facilitate the creation of a vision, goals, and objectives with which to draft a business plan, which then must be implemented and monitored.

Citizen input and local control of the economic process are as vitally important to the long-term integrity, solidarity, and stability of a community as the locally owned, diversified businesses themselves are. Because community development seeks to create a unique circumstance in which human and economic sustainability are mutually reinforcing, the economic content of community development must be carefully thought out and skillfully integrated within the social content.

It is therefore improbable that large, absentee-owned businesses are a realistic answer to local employment needs because such owners are seldom committed to the welfare of the community. They are unlikely to purchase local products, hire local people, or contribute to and remain loyal to the community in difficult times. There are, of course, exceptions.

Moreover, the industrial revolution is continuing apace today in traditional societies but under the guise of "economic development." Its evolution has been, and is, toward larger scale, greater specialization and increasing automation, with ever-increasing integration and interdependence, which inevitably leads to ever-greater vulnerability to systemic failure. At the same time, because industrialization gives a higher standard of consumption (=standard of living) to more people than any other mode of production, its universal dominance is well ensconced.[57]

Global industrialization, with its ever-increasing consumption of energy and material goods, infringes on the functional wholeness of a system. Such infringement denotes a shortcoming in relationships, such as that between industry and the environment, between government and the governed, between the industrialized North and the less industrialized South, and hence between the materially rich and the materially poor.[54]

Although we can infringe on functional wholeness as we will, we cannot do so with impunity. There is a cost of adverse results. The question thus becomes: How sustainable is the current notion of economic development?

The evolutionary history of life on Earth is one of species becoming increasingly specialized. With such specialization inevitably came extinction. But underlying the norm was a foundation of generalists capable of merging themselves into a remarkably wide range of environmental conditions. These generalists were consequently far less subject to either speciation or extinction. For humanity, the lesson is clear. If we want our societies to survive, we must once again become flexible, adaptable generalists and build our societies accordingly. The threat is to contemporary societies. The challenge is to contemporary people. With this in mind, I am going to share a few of Mohandas K. Gandhi's ideas on economic development, ideas that to me make impeccably good sense.[58]

Gandhi looked forward to economic development, but he wanted to "prevent our villages from catching the infection of industrialism." He saw that industrialism led inevitably to the linear-visioned, unrestrained pursuit of material goods, which destroyed a person's purpose in life and all too often led to spiritual bankruptcy.

Gandhi realized that the economy, like the spinning wheel (the symbol of India's bid for freedom), was an organic whole, and that if economic growth was to take place, it must be in harmony with all aspects of the society, especially local communities. It must be holistic. If India tried to develop too rapidly, which in our narrow Western sense means adopting industrial specialization, it would become linear in thinking and shortsighted in vision, getting out of kilter with itself, and negative consequences would follow.

Gandhi saw that if one did not tackle the major problems of the rural economy concomitantly with industrialization, industrial specialization would get too far ahead of agriculture and would cause the latter to grind to a halt. This has already happened to some degree in a number of nonindustrialized countries. "In the first decade of independence," observed *To the Point International* magazine, "not a single Black African country gave priority to agricultural investment, and expenditure on this sector represented only a tiny fraction of the total government disbursement. When industrialization did not deliver the goods...there was no agricultural base to fall back on."[58]

The same thing happened when I was working in Nepal in 1966–1967. The U.S. AID Mission's direction was to bring forestry into the modern era by U.S. standards, ones inappropriate to the Nepalese culture and to the nature of the Nepalese forests. In reality, therefore, we helped the Nepalese to cut their trees and destroy their forests for decades, or centuries, or perhaps forever.

Nation after nonindustrial nation finds itself in this trap. Having committed themselves to rapid linear industrialization, they squander their foreign exchange and their national natural resources on schemes of development unsuitable for their environments or for their cultures. When such schemes fail, they are left broke and hungry, often with a severely damaged environment—the legacy of travel in the "fast lane."

At this point, these nonindustrialized countries are left with only two options. One is to apply for assistance from the United States or some other specialized, industrialized nation or from institutions like the United Nations, the International Monetary Fund, or the World Bank.

If such funding agencies as the World Bank are really interested in serving their customers, they need to place a high priority on learning the language of ecosystems that operate in various degrees of naturalness. Unfortunately, because they fail to do this, the economic relief they proffer comes at a supremely high cost—the loss of self-direction, self-esteem, and all too often the loss of potential ecological sustainability for the indebted nation.

Wendell Berry puts it succinctly: "Communists and capitalists are alike in their contempt for country people, country life, and country places. They have exploited the countryside with equal greed and disregard. They are alike even in their plea that it is right to damage the present in order to make 'a better future.'"[59]

Ecological sustainability is thus the last consideration money-lending institutions are likely to address. After all, they are, by their very nature, out to make as much money as possible. All over the world, the multinational, capitalist-industrial elite extend their reach, searching for trees, animals, minerals, medicines, fossil fuels, or cheap labor that can be converted to profitable commodities. The products of this global economic system are then sold to ever-larger masses of "consumers" in all corners of the globe. The production-consumption tentacles of this industrial growth monster reach into every home in every community worldwide in the form of satellite dishes and television sets.[49]

The second choice of these nonindustrialized nations is to exercise patience and adopt Gandhi's program of building, or rebuilding, self-sufficiency and self-esteem from the bottom up, beginning in the local communities.

For whatever reasons, almost all nonindustrialized nations have chosen and continue to choose rapid industrial specialization. They simply trade in political imperialism for economic imperialism. India, for example, made this mistake when, under Nehru, it turned its back on Gandhi's program and embraced rapid industrial specialization. Today, India has little ability to distribute anything to its impoverished masses from its relatively small but highly advanced technical sector.

In Gandhi's view, if the individual, then the local village (community), and then the nation were brought step by step to economic self-sufficiency, it would be possible to attain and retain true political freedom for the whole. If, however, one was in a hurry to industrialize and became economically dependent on another nation, it was at the expense of one's own liberty and freedom of choice. In order to protect its investment, for example, the World Bank has frequently dictated domestic environmental policy as a precondition for approving a loan and has then commanded a supervisory role in the recipient's economy.

A nonindustrialized nation is thus exposed to grave risks when it opts to enter the international economic system. Once it has become reasonably integrated into the system, it may find that it has unwittingly imported such problems as inflation, the effect of foreign recessions, uncertainties in the price and supply of oil, sudden unemployment, or employment skewed toward the desires of foreign markets—

to the detriment of its own economy. Now the leadership of the nonindustrialized nation is subject to even greater influence by foreign nations, and its economy is even more dependent on the debt-ridden international system, which seems to be on the verge of collapse.

The underlying problem is that today's world leaders want to build from the top down, but without a solid, holistic foundation of human dignity based on social/environmental sustainability. This top-downness of industrial enterprise is propelled by the notion that human intelligence can transcend material limits and thus serve the powerful economic elite.

Although it is possible for some people to become so enthralled with this notion, or so dependent on the enterprise, that they see beauty in huge, belching smokestacks, dams in rivers, seemingly endless power lines, and dumps of toxic wastes that come with it, for the majority of humanity the crude, destructive, inequitable aspects of industry remain glaringly obvious. They are too often accepted, however, and even deemed necessary as the price we pay for material progress.

Gandhi's plan for world order, on the other hand, is predicated on the voluntary cooperation and coordination of friendly states reaching out to one another with dignity, reaching out for mutual benefit in such a way that they can approach the same goal from different, even opposite, directions. Perhaps this is what President Woodrow Wilson meant when he said, "Friendship is the only cement that will ever hold the world together."

Put another way, if we in fact act as one another's keepers, society's internal guiding system can function properly and ensure that all the social components remain in harmony with one another. But if an external force, such as imperialism, or an internal force, such as premature industrial specialization, disrupts the balance, such debilitating problems as urban drift, unemployment, hunger, crime, drugs, child and spousal abuse and abandonment, civil disorders, terrorism, and general violence become the norm.

Although the specialized, industrialized nations seem to have solved the problems of agriculture and mass production, they lack moral fiber and courage. They are rapidly despoiling the environment for short-term economic gain. And the inequitable distribution of wealth that is due to monopoly capitalism and the now collapsed communism threatens to plunge the global society into chaos. But through it all, we have a choice, because Gandhi's view of world order is built on a foundation of human dignity and harmony both within itself and with the environment on which human society depends.

Thus Gandhi would choose an ecologically sustainable environment that honors the cycles of Nature for the lasting benefit of everyone through the long haul, which means developing technology that is ecologically oriented and energy efficient. Instead, we seem mired in a glut of quick, short-term profits for the immediate benefit of a tiny, powerful, linearly thinking, self-centered economic elite. This situation bears out Eduardo Galeano's observation that, "Massive misery is the price poor countries must pay so that six percent of the world population may consume with impunity half of the earth's generated wealth."

THE ROLE OF LOCAL GOVERNMENT

A prerequisite for sustainable development in a local community is that it must be inclusive, relating all relevant disciplines and special professions from all walks of life. Setting a good example is one of the most important functions of any local government involved in implementing the principles and practices of sustainable community development. Leading by example—breaking down bureaucratic barriers through interdisciplinary crossing of departmental lines, recycling and buying recycled goods, car pooling, providing day care, and flexible working hours—increases not only the capacity of a government to govern but also its effectiveness *and* efficiency.

It is thus important for governments to both identify departmental and community links concerning mutually interrelated issues and to bring all people affected to the table in an effort to collectively resolve shared problems, which means dealing with human diversity. Understanding and accepting diversity allows us to acknowledge that each of us has a need to be needed, to contribute in some way. It also enables us to begin admitting that we do not and cannot know or do everything and that we must rely on the strengths of others with complete trust.

Diversity of thought, culture, and expertise thus allows all persons to contribute to the development process in a special way, making their unique gift a part of the effort necessary to create a sustainable local community. Accepting diversity helps us to understand the need each person has for equality, identity, and opportunity in the process. Recognizing diversity gives us all a chance to provide meaning, fulfillment, purpose, and a gift of our talents to our community and future generations.

Assuming people accept the notion of diversity, what is it that they most want from the development process? People want the most effective, productive, and rewarding way of working together to achieve a common end. They want the process and the relationships forged therein to meet their personal needs for belonging, meaningful contribution, having the opportunity to make a commitment to a special place—their community, having the opportunity for personal growth, and having the ability to exert reasonable control over their destinies.

Increasing Local Adaptability

Societies around the globe are in the throes of change, some because of diminishing resources and others because of social upheaval, but all are losing control over their destinies. Regardless of the cause, sustainable local community development provides vision, planning, and direction in times of crisis as well as in times of peace.

The focus of the local government under the auspices of sustainable community development must be on balancing the ability of a community to meet its own needs while maintaining relative economic stability as outside markets fluctuate. This must be done while simultaneously protecting the ability of future generations to meet their needs in the same area.

Sustainable community development works to maintain a dynamic equilibrium through consciously directed, systemic, self-reinforcing feedback loops. It offers a process that can mobilize citizens to direct information toward long-term community sustainability, which in some measure equates with the economic stability of a local community.

Sustainable community development increases the adaptability of a community by creating and maintaining a diversified social and economic base with local shared ownership and access to basic human services. Community adaptability, and therefore stability, is based on its ability to meet the majority of its own needs within itself instead of being dependent on outside resources. This means, however, that the adaptability of a community also encompasses the ecological integrity of its surrounding landscape.

Improving Citizen Participation

A vitally important component of sustainable community development is local citizen participation in planning, implementing, and monitoring programs, policies, and projects. The goal is to improve the quality of popular participation instead of merely its quantity.

Sustainable community development is based on the assumption that the best ideas usually come from the people, not the policymakers. Therefore, active participation in a local community is necessary to direct the process, which means, for example, taking part in citizen administrative boards, in town meetings, and in local grass roots activities.

As a process, sustainable community development exposes citizens to the ramifications of their thoughts and actions on others, their local environment, and the surrounding landscape, as well as motivating and organizing people to direct change within the context of a shared vision for their collective future. Its aim is for citizens to control the developmental process by feeding ideas and information to the governing body through self-empowered organizations.

People want the most effective development process possible, one that is honestly used through participation in a truly democratic way. Participative development must begin with a firm belief in the potential of people. It arises both out of a leader's heart and his or her personal commitment to people and out of the heart of the democratic principle: the right to an open, accessible process; the right and duty to influence decision making; the right and duty to understand the results; and the duty to be accountable for those results.

To accomplish participative development, leaders must create and maintain emotionally safe environments within which people can develop quality relationships with one another. Creating such an environment requires at least six things: (1) respect for one another; (2) understanding and accepting that what people believe precedes policy and practice; (3) agreement on the rights of participation in and access to the planning process; (4) understanding that most people work as volunteers and need personal covenants, not legal contracts; (5) understanding that relationships count more than structure because people—not structures—build trust; and (6) protecting the process against capture by self-serving financial interests.

The development committee's needs are best met by meeting the needs of its individuals. If this is done, development can be productive, rewarding, meaningful, maturing, enriching, fulfilling, healing, and joyful. Participative development is one of the greatest privileges in our democracy and one of our greatest responsibilities.

Nevertheless, the creative development process is difficult to handle because in such a process almost everyone, at different times and in various ways, plays four roles: one as creator, another as implementor, a third as temporary leader with a specific expertise demanded by a given circumstance, and finally as follower, supporter, and helper.

Although implementation is often as creative as the questions to which it is responding, it is at this very point that leaders and managers may find it most difficult to be open to the influence of others. Nevertheless, by conceiving a shared vision and pursuing it together, a local community's problems of cultural adaptability and sustainable development can be resolved, and the community members may simultaneously and fundamentally alter their concept of adaptability, sustainability, and development. But this requires "joint ownership" of the development process.

The heart of sustainable development is joint ownership of the process for each person involved. Because owners cannot walk away from their concerns, everyone's accountability begins to change. Ownership demands increasing maturity on everyone's part, which is probably best expressed in a continually rising level of literacy: participative literacy, ownership literacy, sustainable development literacy, and so on. And ownership demands a commitment to be as informed as possible about the whole.

Joint ownership is an intimate, personal experience in that each person commits himself or herself not only to the process but also to the outcome. One's beliefs are connected to the intimacy of one's experience and come before and have primacy over policies, standards, or practices. This intimate, personal commitment to the development process affects one's accountability and draws out one's personal authenticity.

No development process can amount to anything without the people who make it what it is. It is initially what the people are and finally what the people become. People do not grow by knowing all the answers; they grow by living with the questions and their possibilities. The art of working together thus lies in how people deal with change, how they deal with conflict, and how they reach their potential.

The intimacy of ownership arises from translating personal and community values into a plan for a sustainable future that seeks its excellence in a search for truth, wisdom, justice, and knowledge—all tempered with intuition, compassion, and mercy. The people of a community must therefore make a covenant, a promise with one another: to honor and protect the sacred nature of their relationships so that each may reflect unity, grace, poise, creativity, and justice. If they base decisions on the intrinsic value of human diversity, and if they base decisions on the notion that every person brings a unique offering to the development process, then inclusivity will be the only path open to them.

Including people—really including people—in the development process means helping them to understand the process, their place within it, and their accountability for the outcome. It means giving others the chance to do their best according to the diversity of their gifts, which is fundamental to the equality that environmental justice requires and democracy inspires. Finally, a community must be committed to using wisely and responsibly its environment and its finite resources, which means a conscious, sustainable, reciprocal relationship between the local community and its surrounding landscape.

To create the desired change, however, it is essential that all affected groups in the community be involved in the process and trained in the skills of leadership. It is further necessary that the people responsible for a local program, policy, or project be involved in its creation and monitoring to increase the probability of a successful outcome. If not, a political problem arises because sustainable community development is initially site specific, and that, in my experience, inevitably brings up either turf struggles or a blatant denial of both responsibility and accountability by passing the buck.

I spent over two years on a citizen's committee to advise my home county on environmental issues. During that time, the city and county did not coordinate between the city and its surrounding landscape, especially with respect to water for the future (which I will discuss later). In turn, when things got uncomfortable, the county insisted that it did not have the jurisdiction and therefore neither the responsibility nor accountability. Those belonged to the state because the land use plan had been carried out at the state level.

Such top-down planning does not work because communities have no vested interest in doing what they feel will benefit the state by benefiting other communities at their perceived momentary expense. The counties, in turn, are concerned about their own interests and turf. Thus, little gets done with sufficient forethought to be of real long-term social/environmental benefit to the future of the community, county, or state. Oh yes, there are the interminable meetings, but little significant action and even less accountability.

As long as the majority of the people in a community, county, state, or nation are predominantly self-centered, and thus myopic, each and every level of government must see a clear—and often immediately personal—advantage before cooperation and coordination become a reality. This is important, because to cooperate and coordinate implies the willing acceptance of both responsibility and accountability, which most people avoid whenever possible. Whatever people do, it is clearly demonstrated by the self-reinforcing information feedback loops.

INFORMATION FEEDBACK LOOPS

People's values, belief in process, and the empowerment to act and collectively resolve problems is the first component of sustainable community development. Education that allows people to learn of their connection to social/environmental problems, both local and global, is the second part of the mechanism. Teaching participants how to plan strategically is the third part, and a sound working knowledge and practice of the democratic system of government within which to fit the first three parts is the final, all-encompassing piece. It is all encompassing because its processes of public representation funnel people's knowledge, feelings, requirements, desires, and concerns into an informational feedback loop that directs change. In this manner, sustainable community development can have a significant effect in directing societal change toward environmental sustainability.

Sustainable community development can instill a sense of purpose and a sense of belonging that are defined and maintained by a local community within its environmental context. It does this by integrating all aspects of society in working toward a dynamic balance of sustainable outcomes.

Such balance can only be maintained if information is fed back into the system in a way that fosters new questions and new practices appropriate for changing circumstances while simultaneously discarding only inappropriate old questions and old ways. Sustainable community development is based on information feedback, not insanity, which is trying the same old thing over and over while each time expecting a different and more desirable outcome.

Sustainable community development creates a mechanism for information to feed through the political system and direct change toward a dynamic equilibrium between the community and its environment. In the case of adjacent communities, a collective mechanism of information feedback is also essential if sharing the same landscape and its common products, such as water, is going to be just and sustainable.

Wise and sustainable development requires our total conscious presence in the present. It also requires that decisions be based on true human/environmental indicators, such as feelings balanced between logic and intuition, a balance between questions based on social value and questions in search of scientific understanding, a balance between social and environmental necessities, self-worth, responsibility in the present for the future, community (including economic)/ecological adaptability, and so on.

This may sound easy, but it is extremely demanding in terms of concentration and energy. The most difficult part of developing community/environmental sustainability is that we will all have to let go of some of our old, cherished beliefs and desires, such as the long-held simplistic notion that economic development is the sum total of community development.

And we will have to face new challenges as we alter the ecosystem within which we live and participate, thereby making it ever-more fragile and our relationship to it ever-more labor intensive and energy demanding. Consider, for example, that in remote times, the nomadic peoples, driven from their homes by scarcity of game and/or impoverished soil, migrated sometimes great distances in search of food and water. In so wandering, the peoples of old escaped many of the diseases that afflict modern society.[60]

But when people began settling into more permanent communities, their health problems increased immediately. As wanderers, they had left their refuse behind and moved into healthier environments, but as city builders, they piled their refuse in the outskirts of the villages and towns, and there was born a great part of the sickness that still plagues society.[60]

Although the problems of sanitation have been largely solved in the rural communities and large cities of the wealthy industrialized nations, such is not the case in much of the world. And even a wealthy industrialized nation like the United States still has its stubborn problems. I say stubborn because they are ignored as much as possible by that part of the populace interested more in personal financial gain than in human/environmental welfare, present or future. A few of these problems, which challenge the possibility of sustainable community development, but by no means all of them, are discussed in the next chapter.

SOME MISSING PIECES OF SUSTAINABLE COMMUNITY DEVELOPMENT

4

C ommunity development as a tool does not in and of itself promote sustainability. No one, as already stated, is safe from the many environmental and social problems that threaten our planet and our health, but we are not all endangered equally. In light of this thought, consider today's notion that local cultures must change and adapt to national and international influences—influences that promote and produce nonsustainable activities and results.

For example, a timber company exporting whole logs from local forests to Japan or a sawmill owner installing state-of-the-art equipment to reduce the number of employees while increasing production of milled lumber for shipment to Japan may help short-term economic stability but simultaneously fosters a local community's dependency on outside markets and their influences. Through such activities, communities become increasingly dependent on external markets and political forces over which they have decreasing amounts of control. Under such circumstances, the importance of the local people and their shared place declines in the face of outside influences. Unfortunately, this is happening in virtually all societies because of expansionist economic cookie-cutter ideas about progress and development.

Community development is only as sustainable as the values, perceptions, and questions that drive it. Here one might ask what community and sustainability have in common. Their commonality is that both must abide by, maintain, or enhance the local culture over time.

As previously discussed, local community development is a process of educating, organizing, and acting, which empowers local people to influence their destiny through self-determined democratic governance. What happens if we now add the notion of sustainability to the equation?

Bear in mind that there is little real chance of moving forward without communities of people and dynamic leaders who are genuinely committed to sustainable types of change. The search for quick fixes usually results in superficial changes with respect to some problematic symptom but leaves untouched the deeper cause of the symptom. The nature of the commitment, which encompasses the need for change in the larger world, uses our local communities as vehicles for bringing about such change, which necessitates the willingness to learn.

Real learning—the remembrance of things forgotten and the development of things new—occurs in a continuous cycle over time. Learning encompasses theoretical conceptualization, practical conceptualization, action, and reflection, including equally the realm of the intellect and that of the intuition. Real learning is important because overemphasis on action, one part of which is competition, simply reinforces our fixation on short-term quantifiable results. Our overemphasis on action precludes the required discipline of reflection, which is a persistent practice of deeper learning that often produces measurable consequences for long periods of time.

As previously stated, many of today's problems resulted from yesterday's solutions, and many of today's solutions are destined to become tomorrow's problems. This simply means that our quick-fix social trance blinds us because we insist on little ideas that promote fast results regardless of what happens to the system itself. What society really needs are big, systemic ideas that both promote and safeguard social/environmental sustainability.

Where, asked the late publisher Robert Rodale, are the big ideas, those that change the world? They probably lie unrecognized in everyday life because our culture lacks sufficient free spaces for general thought.

SUSTAINABLE COMMUNITY DEVELOPMENT AS A BIG IDEA

A big idea, according to Bob Rodale, has the following characteristics: (1) it must be generally useful in good ways, (2) it must appeal to generalists and give them an advantage over specialists, (3) it must exist in both an abstract and a practical sense, (4) it must be of some interest at all levels of human concern, (5) it must be geographically and culturally viable over extensive areas, (6) it must encompass a multitude of academic disciplines, and (7) it must have a life over an extended period of time.[41]

Let's see if sustainable community development could be such a big idea. Local community development is a tool that is subject to the beliefs of those who use it. The context—the shared vision—that guides the process of local community development determines the outcome. Sustainable community development therefore places local community development within the context of sustainability.

Development as a process means building capacity, and community development means building the capacity of people to work collectively in addressing their common interests in the local society.[61] Sustainable community development means building the capacity of people to work collectively in addressing their common interests in the local society within the context of sustainability—that which is sustainable biologically, culturally, and economically.

Sustainable community development is therefore a community-directed process of development that is based on: (1) transcendent human values of love, trust, respect, wonder, humility, and compassion; (2) active learning, which is a balance between the intellect and intuition, between the abstract and the concrete, between action and reflection; (3) sharing generated through communication, cooperation, and coordination; (4) a capacity to understand and work with and within the flow of life as a fluid system, recognizing, understanding, and accepting the significance of relationships; (5) patience in seeking an understanding of a fundamental issue rather than applying band-aid-like quick fixes to problematic symptoms; (6) consciously integrating the learning space into the working space into a continual cycle of theory, experimentation, action, and reflection; and (7) a shared societal vision that is grounded in long-term sustainability, both culturally and environmentally.

Sustainable community development seems to fit all of Rodale's requirements. It also helps people to understand that life is not condensable, that any model is an operational simplification, a working hypothesis that is always ready for and in need of improvement. With the understanding that there are neither shortcuts nor concrete facts, communication functions as a connective tool for and through sharing invention, cooperation, and coordination.

When people speak from their hearts and unite through active listening, they produce tremendous power to invent new realities and bring them into being through collective actions. Thus, while today's environmental users with special interests will not all be around in the next century, all of the environmental uses and all of the sustainable local communities can be if we achieve sustainable development at the local community level as a "big idea."

CHANGE AS A LOCAL CREATIVE PROCESS

The time for sustainable development in local communities is fast approaching because citizens all over the United States, indeed the world, are realizing that *they* must take the lead in addressing their own social/environmental problems rather than waiting for their leaders to limply take the initiative. As a society, however, we have largely chosen to waste time pointing fingers at universities, agencies, and Congress, rather than risk taking charge of our own destinies.

Nevertheless, many people, regardless of their particular walks in life, are awakening to the need for change. People are realizing that their democratic society is changing and will continue to change without them unless they begin actively participating in the process of governance. They are awakening to their own powers of self-support and self-management. Given their own tools, local communities are more effective in defining and meeting their own needs, at lower costs, than are state and federal governments or private service providers, and so begins a populist movement.

A populist movement reaches its peak whenever the fundamental institutions of a nation (large corporations and organized religions, as well as the governments, from municipalities to nations) consolidate vast powers but collapse inwardly from corruption and collusion among policy and business elites. This often results in bureaucratic paralysis and lack of accountability. Real issues are then ignored or weakly addressed through a series of political charades, because while science and media thrive on controversy, politics is paralyzed by it.

For example, current public environmental debates are so politicized that we become preoccupied with relatively minor details, which causes us to run around looking for knowledge while we are drowning in information. This happens whenever we fail to proceed from a basic frame of reference that allows us to focus on the fundamental issues and real reforms without getting lost in a confusion of isolated and isolating details.

Although most people admit that farsighted, far-reaching reforms are urgently needed, the shallowness of political dialogue causes public cynicism and feelings of being manipulated. Thus, in an atmosphere of growing crises, local communities discover that they must resolve their own problems, which is the very basis of sustainable community development.

In the community development process, values are shared and learned. The question is not whether we transmit values, because we do; the question is what values we transmit and how.[61] The values that need to be learned and transmitted in order to move toward sustainability are those of a unified systemic world view: values of love, dignity, relationship, equality, compassion, justice, respect, democracy, responsibility, and process.

These values, which underlie sustainable development, can be transmitted and implemented through education, self-actualization, self-help, and collective action—the process of local community development. There are, however, no teachers with correct answers; there are only guides, such as facilitators, with different areas of expertise and experience who may help along the way.

Through community development, a process is created that provides a suite of applied strategies for bringing about change in such a way that tools are available for public adoption and use. By placing the community development process within the context of sustainability—shifting from the expansionist economic world view with its reductionistic mechanical foundation to a unified systems way of thinking—one builds a description and analysis of change as a process based on sustainable values and aspirations.

Placing the process of change within the realm of sustainability directs local communities toward a better understanding of the effects of individual and collective participation within the community itself and its immediate environment within its surrounding landscape. It also encourages people to change from nonsustainable activities to sustainable ones.

A prerequisite for working toward sustainable outcomes from decision making and developmental action is to increase people's awareness of: (1) their connection to one another within their local community, (2) their communal connection to their immediate environment and its surrounding landscape, and (3) their community's place within the larger regional, state, national, and global levels of society. As people come to understand and feel their role as a contributor to the many issues that we as a society face, the need for action and change will become increasingly powerful.

People's sense of empowerment and belief in their potential to resolve problems is crucial to sustainable community development. People are a powerful catalyst for change both as activists within their local communities and as examples in changing their behaviors to promote more sustainable lifestyles. Such change relies on creativity, which is fragile and easily stifled. Each person's creativity must therefore be encouraged, developed, and protected if sustainability is to be a viable part of community development.

When people are inspired by their own interests and enjoyment, there is a better chance they will explore unlikely paths, take risks, find that employment and income can exist within sustainable practices, and produce something that in the end is useful. Motivation is internal to people themselves. Thus, while people must motivate themselves, they can be helped to analyze, understand, and use their own experiences to new and greater ends. This brings up a Chinese proverb: "I hear and I forget; I see and I remember; I do and I understand." In this sense, people are not trained, although training in formal skills may be valuable; they are liberated and train themselves, using other people as examples and in turn providing examples to still other people.

When goals are imposed on them, however, or when they are goaded by fear of firing, creativity withers. Sustainable community development therefore depends on intrinsic motivation, which is conducive to creativity.

Although our larger social system is designed to insist on conformity, to go along with mass thinking, local sustainable community development by its very nature is designed as an advocate and protector of the freedom and space necessary for creativity to flourish. The single most important component of creativity is freedom—the power to decide what to do, how to do it, and when to do it, a sense of control over one's own ideas and work. "Genius," according to inventor Thomas Alva Edison, "is 1 percent inspiration and 99 percent perspiration." His observation seems to be true. To physicist Albert Einstein, what genius needs most is the freedom of pursuit.

Because many local issues can and must be addressed simultaneously through the process of sustainable community development, it is a potentially powerful strategy for change. By addressing the needs and concerns of both individuals and community groups, it increases the solidarity and adaptability of a local community, which means that the issues on which a community focuses become increasingly centered within the context of long-term sustainability.

Long-term social/environmental sustainability is a consciously directed process within the democratic system; it is the development phase of a sustainable community. The development process begins with the creation of a shared vision and goals (see *Resolving Environmental Conflict*[25]). Grasping a shared vision creates momentum, which demands a well-conceived strategy to achieve that vision. Such a strategy must include clearly communicated directions and workable means that enable everyone to participate initially in creating the development plans and later in being publicly accountable for achieving them.

The development process, however it evolves, is important because only through it can people participate in a shared vision of the future. If we cannot share a vision for the future toward which to build, then we are, as Professor Aldo Leopold wrote in 1933, confronted by a contradiction: "To build a better motor we tap the uttermost powers of the human brain; to build a better countryside we throw dice."

Local people who empower themselves to work together in tapping the utmost powers of the mind, intuition, and experience in developing their own sustainable community will reap the following benefits:

1. A defined course of action, which helps ensure that the selected course has a good chance of success.
2. A process that serves as the foundation on which all community activities are based and, as such, must result in answers to what,

where, when, why, and how actions are to be taken and who will conduct the actions for whom. If these questions are answered in a manner satisfactory to all, the chance of a destructive conflict is greatly reduced, but if perchance conflict arises, it can usually be resolved, often within the context of sustainable community development.

3. A well-conceived plan that allows those responsible to determine what they are responsible for and provides people with the opportunity to gain a clear insight with respect to their specific tasks in relationship to the function of a community as a whole.

4. A process that helps a particular group of people communicate to others that the group is thoughtful in what it is doing and stands a good chance of accomplishing its stated purpose.

5. A process that will aid in monitoring and evaluating a community and its achievements.

6. A periodic evaluation of a group's progress toward meeting its vision, goals, and objectives identified in the plan, which is critical for evaluating whether it is providing the promised services to its customers and supporters. This step is essential for the health and growth of any community or organization within a community.

7. A vehicle through which the collective long-range (100+-year) vision of the people involved with a community can be realized. Planning, which is looking at options and solutions, helps people focus their energy on their vision, goals, and objectives and thereby helps a community achieve maximum utilization of its human talents and financial resources.

8. A process that helps people to influence what the present is and the future might be for the benefit of both today's citizens and tomorrow's generations.

In addition to all the goals, parameters, and legal requirements embedded in the planning process, it is fundamental that leaders endorse the concept of persons, which begins with recognition, understanding, and acceptance of people's diversity in their creative gifts, talents, and skills. These creative gifts, talents, and skills will also be needed to design lifestyles that are in sustainable harmony with the environment, which affects the sustainable harmony of the whole world.

THINK GLOBALLY, ACT LOCALLY

"Think globally, act locally" is a critical concept because life in any one place, such as a human community, must and can exist only within the context of a healthy and sustainable biosphere, the area in which the living component of the world exists. The biosphere, in turn, is sandwiched between the lithosphere (the Earth's geological mantle) and the atmosphere (the air, with allows life as we know it to exist).

Thus, while "think globally, act locally" says a lot if one has traveled abroad and worked in other cultures, it is very much an abstraction to people who have been content to stay at home. Nevertheless, there are many ways to put the notion of "think globally, act locally" into action within a community. Let's look at two examples.

Controlling the Human Population: A Matter of Gender Equality

We have been warned for decades that the human species is overpopulating the Earth. Yet our population explodes and the usable portion of the Earth per individual shrinks, as does the allotted proportion of its resources, all of which become more quickly limiting when abused. We have tried many things to remedy this situation: education, birth control, feeding the hungry, shipping industrial technology to poor nations, and so on. In my opinion, however, we have not addressed the primary cause of overpopulation: the inequality between men and women.

Women have long been dominated by men. Through such domination, women are physically forced to produce most of the world's food yet are allowed to own but an infinitesimal part of the land. Women have had only one way to be uniquely valued by men— having babies. And since most societies are patriarchal, some to an extreme, having babies—primarily sons—also equates with a woman's social value.

Regardless of where I have traveled, I have found that women who have a good education have fewer children and have them later in life. Education affords increased opportunities and a variety of ways to be valued. I submit, therefore, that if we are serious about controlling the human population, women must have an equal voice in all decisions and unequivocal access to opportunities for self- and social valuation.

On the surface, this means such things as equal opportunity for education, jobs, and pay and equal pay for equal work. At its root, this means changing the male attitude of superiority toward women—a difficult task but a vitally necessary one.

Equality is also a basic tenet of democracy, a tenet that is still denied women more than 200 years after the United States of America was founded on the ideal of human equality. If gender inequality is a community problem, it is a regional, national, and global problem. Therefore, gender equality within a community begins to create gender equality on a regional, national, and global basis. After all, the global society is composed of communities, each of which adds its unique touch, each of which adds its influence to the whole.

Controlling the human population begins here, in our local community. It begins within each family, among neighbors, among acquaintances, among sales people and customers, among bosses and employees, and among elected officials who represent the people. The community that practices *real* gender equality within itself begins to change the world and to control the human population. If you doubt me—ask the women!

Healthy Rivers and Oceans Are a Matter of How We Care for Local Ditches

Did you ever think about a ditch? Just an ordinary roadside ditch? Most people probably do not even notice them, much less think about them. Nevertheless, there was actually a time in the world before ditches, a time when water itself decided where humanity would dwell. Then the ditch was invented, and that changed everything.

Somewhere in time, that first ditch became a conscious thought that translated into a conscious act. As the one ditch became the many ditches, humanity and plants and animals moved into areas previously uninhabitable by those who needed water in close proximity; thus the human sense of place was expanded.

The first ditch irrevocably altered humanity's sense of itself, its sense of society, and its ability to manipulate Nature. Ditches gave rise to agriculture and eventually led to such feats of engineering as the Panama and Suez canals, each of which physically connects one ocean with another.

Today, however, the ecological health of roadside ditches reflects the state of our environment, and the ditches I see are too often but conduits for humanity's sewage and toxic wastes, linear roadside dumps for society's materialistic offal. And they are an integral but unrecognized and ignored part of the stream-order continuum.

Although the stream-order continuum is a concept devised for streams, I find many of the same processes in ditches, and so I think of it as the "stream/ditch-order continuum." The stream/ditch-order continuum operates on a simple premise: Streams are Nature's arterial system of the land, and ditches create culture's arterial system. As such, they form a continuum or spectrum of physical environments, with associated aquatic and terrestrial plant and animal communities, as a longitudinally connected part of the ecosystem in which downstream processes are linked to upstream processes.

The idea of the stream/ditch continuum begins with the smallest stream or ditch and ends at the ocean. The concept centers around the resources of available food for the animals inhabiting the continuum, which range from invertebrates to fish, amphibians, reptiles, birds, and mammals.

As organic material floats downhill from its source to the sea, it becomes smaller, while the volume of water carrying it becomes larger. Small streams thus feed larger streams and larger streams feed rivers with partially processed organic matter, the amount of which becomes progressively smaller the farther down the continuum of the river system it goes. The same is true for ditches.

This is how the system works: A first-order stream is the smallest undivided waterway or headwater, a description that fits most ditches. Where two first-order streams join, they enlarge as a second-order stream, again a description that fits ditches. Where two second-order streams come together, they enlarge as a third-order stream and so on.

The concept of stream order is based on the size of the stream (the cumulative volume of water) and not just on which stream of what order joins with another stream of a given order. For example, a first-order stream (or ditch) can either join with another first-order stream (or ditch) to form a second-order stream (or ditch) or it can enter directly into a second-, third-, fourth-, fifth-, or even larger order stream. The same is true of a second-order stream, a third-order stream, and so on.

In addition, the stream/ditch order influences the role played by streamside and ditchside vegetation in controlling water temperature, stabilizing banks, and producing food. Streamside vegetation is also the primary source of large organic debris, such as tree stems at least eight inches in diameter with their rootwads attached or tree branches greater than eight inches in diameter. Ditches, on the other hand, are usually stripped of their trees long before they mature.

Forests adjacent to streams supply wood (stems, rootwads, and large branches from trees) while ditchside vegetation supplies grasses, herbs, and occasionally branches from shrubs, but rarely from trees. Erosion also contributes organic material to the stream or ditch.

Wood in streams increases the diversity of habitats by forming dams and their attendant pools and by protecting backwater areas. Wood also provides nutrients and a variety of foundations for biological activity, and it both dissipates the energy of the water and traps its sediments. All these functions in ditches are usually performed by nonwoody vegetation.

Processing the organic debris entering the aquatic system includes digestion by bacteria, fungi, and insects and physical abrasion against such things as the stream bottom and its boulders or the ditch bottom and its pebbles. In all cases, debris is continually broken into smaller pieces, which makes the particles increasingly susceptible to microbial consumption.

The amount of different kinds of organic matter processed in a reach of stream or ditch (the stretch of water visible between two bends in a channel, be it a ditch, stream, or river) depends on the quality and the quantity of nutrients in the material and on the stream or ditch's capacity to hold fine particles long enough to complete their processing. The debris may be fully utilized by the biotic community within a reach of stream or ditch or it may be exported downstream.

Debris moves fastest through the system during high water and is not thoroughly processed at any one spot. The same is true in streams and ditches that do not have a sufficient number of instream or inditch obstacles to slow the water and act as areas of deposition, sieving the incompletely processed organic material out of the current so its organic breakdown can be completed. Thus, small streams feed larger streams and larger streams feed rivers, just as small ditches feed larger ditches, which eventually feed streams and rivers.

As a stream gets larger, its source of food energy is derived more from aquatic algae and less from organic material of terrestrial origin, which my observations suggest is comparable for ditches. The greatest influence of terrestrial vegetation is in first-order streams and ditches, but the most diversity of incoming organic matter and the greatest diversity of habitat is found in third- to fifth-order streams and large rivers with floodplains.

Small, first-order, headwater streams (and, where applicable, ditches) largely determine the type and quality of the downstream habitat. They and second-order streams and ditches are influenced not only by the configuration of surrounding land forms but also by the live and dead vegetation along their channels. This vegetation is called riparian vegetation and interacts in many ways with the stream or ditch.

The canopy of vegetation, when undisturbed, shades the streamside or ditchside. The physical energy of the flowing water is dissipated by wood in stream channels and by grasses, sedges, rushes, and cattails in ditch channels, slowing erosion and fostering the deposition of inorganic and organic debris.

Because these small streams and ditches arise in tiny drainages with a limited capacity to store water, their flow may be scanty or intermittent during late summer and autumn, but during periods of high flows in winter and spring, they can move prodigious amounts of sediment and organic material.

What I have just described is the beneficial aspect of the stream/ditch continuum. There is, however, a sinister side to the ditch portion of this story, a tragically human side.

Remember that ditches form a continuum or spectrum of physical environments (the same as streams) along a longitudinally connected part of the ecosystem in which downstream processes are linked to and influenced by upstream processes. Remember also that the stream/ditch continuum begins with the smallest stream or ditch and ends at the ocean. Little ditches thus feed bigger ditches and bigger ditches eventually feed streams and rivers, which ultimately feed the ocean. Remember further that as organic material (food energy) floats downhill from its source to the sea, it gets smaller (more dilute), while the volume of water carrying it gets larger.

But what happens to the continuum concept when a ditch is polluted? To pollute a ditch means to contaminate it by dumping human garbage into it or by discharging noxious substances into it, both of which in one way or another disrupt biological processes, often by corrupting the integrity of their chemical interactions.

While Nature's organic matter (food energy) is continually diluted the further down the continuum it goes, pollution (especially chemical pollution) is continually concentrated the further down the continuum it goes because it gathers its potency from the discharge of every other contaminated ditch that adds its waters to the passing flow. Thus, with every ditch we pollute, the purity of streams and rivers is to that extent compromised, and the amount of pollution that humanity is dumping into the estuaries and oceans of the world through the stream/ditch continuum is staggering.

I say this not only because I have seen ditches in North America, Europe, Asia, and Africa discharging their foul contents into streams, rivers, estuaries, and oceans but also because of a population of montane voles (meadow mice to most people) I found in 1969 living along a ditch that drained an agricultural field. The voles, whose fur was an abnormally deep yellow when I caught them, lost the yellow with their first molt in the laboratory when fed normal lab chow; those along the ditch, on the other hand, retained their yellow pelage.

No matter how hard I tried, I could find no one in agricultural chemistry at the local university to acknowledge this color deviant, let alone examine it in a effort to find the cause—undoubtedly some agricultural chemical compound. They all turned their backs, even when I presented them with the evidence, the live yellow voles. Thus I learned that chemical pollution in ditches is not visible to the human eye in the flowing of the waters, but it may become visible in the sickening of the environment.

In 1984, as part of a committee called to Washington, D.C. to help the U.S. Congress frame the ecological components of the 1985 Farm Bill, I learned in far greater depth about the incredible nonpoint source chemical pollution of our nation's surface waters (ditches) and groundwaters (aquifers) from today's chemical-intensive agriculture.

How, I wonder, can we learn to care for rivers and oceans if we continually defile the ditches that feed them? The answer is that we cannot!

We must learn to care first and foremost for the humble things in our environment, such as a roadside ditch, before we can learn how to care for the mighty things in our environment, such as a river. Defile the ditch and we defile the river, estuary, and ocean; protect the ditch and we protect the river, estuary, and ocean. Thus is Nature's lesson taught, a lesson that begins when we are children in a community.

How long will it take for society to kill the mighty oceans of the world by poisoning the humble ditches of the land? The answer, in large part, depends on how clean and healthy each community keeps its ditches. Again, the human actions taking place in each and every community reach the whole world and, in the collective, shape the destiny of humanity with respect to the quality of lifestyle that will be possible in the future.

CULTURAL CAPACITY, CARRYING CAPACITY, AND SUSTAINABILITY

Although we may think ourselves wise in our own eyes, we are too often blind to the truth that we neither govern nor manage Nature. We treat Nature wisely or unwisely for good or for ill, but we do not control Nature. We do something to Nature, and Nature responds, and in the response lies the lessons we are loathe to learn—lessons about lifestyle.

Lifestyle is commonly defined as an internally consistent way of life or style of living that reflects the values and attitudes of an individual or a culture. Many in Western society have made lifestyle synonymous with "standard of living," which we practice as a search for ever-increasing material prosperity. If, however, we are to have a viable, sustainable environment as we know it and value it, we must reach beyond the strictly material and see lifestyle as a sense of inner wholeness and harmony derived by living in such a way that the spiritual, environmental, and material aspects of our lives are in balance with the capacity of the land to produce the necessities for that lifestyle.

Whether a given lifestyle is even possible depends on "cultural capacity," a term that is an analogue for "carrying capacity," which is the number of animals that can live in and use a particular landscape without impairing its ability to function in an ecologically specific way. If we want human society to survive the twenty-first century in any sort of dignified manner, we must have the humility to view our own population in terms of local, regional, national, and global carrying capacities, because the quality of life declines in direct proportion to the degree to which the habitat is overpopulated.

If we substitute the idea of cultural capacity for carrying capacity, we have a workable proposition for sustainable community. Cultural capacity is a chosen quality of life that is sustainable without endangering the productive capacity of the environment. The more materially oriented the desired lifestyle of an individual or a community, for example, the more resources are needed to sustain it and the smaller the human population must be per unit area of landscape. Cultural capacity, then, is a balance between the way we want to live, the real quality of our lifestyle and our community, and the number of people an area can support in that lifestyle on a sustainable basis.

Cultural capacity (=quality) of any area will be less than its carrying capacity (=quantity) in the biological sense. Cultural capacity has built into it the prudence of limitations as a margin of safety in the event of such long-term phenomena as global climate change. Carrying capacity, on the other hand, uses the environment to its maximum and lacks a margin of safety for difficult years or unforeseen environmental changes. The long-term environmental risks hidden in the momentary notion of an area's carrying capacity, when exploited to the maximum, have doomed more than one civilization to collapse by destroying the biological sustainability of the surrounding landscape.

Cultural capacity is a workable idea. We can predetermine local and bioregional cultural capacity and adjust our population growth accordingly. (Bioregion is used as a geographically definable area of biological similarities, which is largely self-contained when it comes to a supply of water.) If we choose *not* to balance our desires with the land's capabilities, the depletion of the land will determine the quality of our cultural/community/social experience—our lifestyle.

So far, we have chosen not to balance our desires with the capabilities of the land, because we have equated *desire* and *want* with *need* and *need* with *demand*, as though our desires and wants are somehow necessities, which they are not. We have thus lost sight of ecological reality because we view the world through the ever-greedy eyes of the consummate impossible-to-satisfy consumer.

If, therefore, we desire to maintain a predetermined lifestyle, we must ask new questions: (1) How much of any given resource is it *necessary* to leave intact as a biological reinvestment in the health and continued productivity of the ecosystem? (2) How much of any given resource is *necessary* for us to use if we are to live in the lifestyle of our choice? (3) Do sufficient resources remain, after biological reinvestment, to support our lifestyle of choice, or *must we modify* our lifestyle to meet what the land is capable of sustaining?

Necessity is a proposition very different from the collective "desire, want, need, demand" syndrome; therefore, arguments about the proper cultural capacity revolve around not only what we think we want in a materialistic sense but also what the land can produce in an environmentally sustainable sense. Cultural capacity is a conservative, other-centered concept, given finite resources and well-defined values. By first determining what we want in terms of lifestyle, we may be able to determine not only if the landscape can support our desired lifestyle but also how we must behave with respect to the environment if we are to maintain our desired lifestyle.

Cultural capacity is therefore about limits to the growth of a community both in area and in absolute numbers of people. But I have heard it said that we have no legal method or even the "right" to keep people out of a community by limiting growth if they choose to live there. If this is true, then no community has any real control over its own destiny, because the people who are the community have no control over the size of their population and therefore over the sustainability of either their community or the immediate environment.

I would ask, however, whether it is culturally *necessary* to do such things as entice industrial growth into an area with tax breaks and/or other devices that the average community member is denied or whether it is economic greed. Industrial growth means an increase in the number of jobs, but not necessarily the *quality* of jobs. And it is the quality, not the quantity, of jobs that must be the focus as a measure of economic health.

Beyond some point, enticing industry into an area is to convert the potential of cultural capacity, which is qualitative, into carrying capacity, which is quantitative. Here, one must remember that the more an ecosystem is altered, the more fragile it becomes, the easier it is broken, and the more labor-intensive and energy-demanding the sustainability of its functional processes becomes if the system as a whole is to be sustainable.

Do we, as members of a community in the collective, have a moral or legal right to destroy the livability of the community for all of the future through continual growth and economic expansion that benefit a few in the present while discounting the future? Each community must address these questions because, by its actions, each community is committing the future to live a quality of life that is predetermined by its cultural/economic middens, including how it alters the surrounding landscape.

COMMUNITY WITHIN THE CONTEXT OF LANDSCAPE

The purpose of this section is to help you understand a little about society's reciprocal relationship to habitat, habitat's relationship to climate, climate's relationship to human society, and to give this discussion some sense of both spatial and temporal scale.

What the generations of the future will inherit from us depends, far more than most people realize, on how we humans treat landscapes, both those which immediately surround our communities and those which connect rural communities with one another and with cities. In addition to immediate human alteration of landscapes around communities and the extinction of species due to subsequent losses of habitat, we are today confronted with unprecedented change in the global climate, which may well have social repercussions of staggering magnitude.

For example, according to *Global Biodiversity Assessment*, a United Nations report released on November 14, 1995, 90 percent of the world's current supply of food comes from just 100 species of plants, while thousands of other sources of food and medicine are being ignored or obliterated. The 1,140-page report, which is the result of work by some 1,500 scientists from around the world, goes on to say that those 100 species of plants from which our supply of food comes are constantly being bred to give higher yields and make farming easier. Such intensive breeding—selecting for certain genetic characteristics of yield at the expense of genetic adaptability—is weakening the species' abilities to overcome diseases and/or changes in climate.[62]

Part and parcel of changes in the global climate will be a greatly increased mobility of both habitats and whole ecosystems as they wander across landscapes. While in past millennia, they were free to wander, today they are constrained by human alterations of those same landscapes, such as the sprawl of cities and the spaghetti-like intersecting networks of superhighways. And because of these impediments, whole segments of habitats, such as forests, may be halted in their northward migration, only to meet their demise as the climate becomes too warm too fast for their survival.

Nomadic peoples of the past had to follow the seasonal ripening of plants and the movements of game animals in order to survive. With the advent of agriculture, many of these nomads settled down and developed a sense of place. As the climate changes, however, and habitats within ecosystems or whole ecosystems begin to move northward in latitude or upward in elevation, modern people, village by village, community by community, will either have to move with their habitat or adjust in place to changing conditions, which may be uncomfortable at best.

While in Malaysia, for example, I asked a college professor about the Malaysian climate, which is essentially equatorial. I was told that the climate was hot and hotter, wet and wetter, humid and more humid. He said that the climate over the last decade had become uncomfortably hot, even for him, who was born there. He said that he stayed in air-conditioned buildings as much as possible.

Although it may not seem like it, all environmental effects that I can think of begin at the level of the local community, including alterations in landscapes.

Local Landscape

It is imperative that local communities not only consider but also involve their surrounding landscapes in all aspects of sustainable development because the spatial patterns of topography and vegetation result from complex interactions among physical, biological, and social forces. For example, in the vicinity of Taos, New Mexico, the indigenous Pueblos Indians had long used the surrounding old-growth piñion pine forest for both food (its seeds) and firewood. Then the Spanish invaded the area and also began using old-growth pine for firewood. It became traditional.

Later, the Anglos invaded and the population grew. Since old-growth piñion pine was the best firewood, everyone wanted an equal share. In recent years, however, there has been a great influx of retired Anglos from Texas and Southern California, most wanting their "fair share" of the firewood.

The result of this continual and growing onslaught on the slow-growing, centuries-old piñon pine forest is the potential for its imminent demise, because there is relatively little of it left. Along with the biological decline of the old-growth forest surrounding Taos is the rapidly growing problem of air pollution from the continual increase in wood smoke.

The people of Taos, by ignoring the reciprocal nature of the participation of their community with its surrounding landscape, are seeing the cultural ambiance of the town's setting fading into history with its rapid growth in population, increased cutting of the old-growth piñon forest for firewood, increasing air pollution with the burning of that firewood, and the loss of its ancient forest by clinging to the tradition of cutting it down for firewood.

Thus, while the sustainability of a community is site specific in one sense, in another sense, communities within close proximity to their neighbors affect those neighbors because the landscape and its ecological patterns form a common denominator among adjacent communities. While each community must determine how its proposed actions will affect its immediate surroundings, it must also consider its direct and indirect effects on its neighbors.

In order to address the likely consequences that the proposed actions of a community may create in the larger context of shared landscape patterns, it is imperative to understand that the value of doing so must be built on the aggregate of communal shared visions from the bottom up, from cooperation and coordination among the communities themselves. It cannot be imposed by state authority from the top down. A simple example is the myriad ways in which a community near the headwaters of a river can affect all communities downstream by its actions within the upper water catchment.

Thus, while the government of a state may generally impose a top-down state-wide vision on the communities within its borders, it will be of relatively little consequence if the communities choose not to comply with the dictated values. It is thus the net effect of the collective visions of all the communities as implemented on the ground that actually governs the future patterns of the landscape within the state.

The state's vision, therefore, is *de facto* the collective of the communities' visions because each community in its development must and will change something. It is thus necessary that change be consciously understood, accepted, and directed as an ongoing process of hidden opportunities yet to be discovered if sustainability in community development is to be achieved. (For a discussion of the various kinds and scales of change, see *Global Imperative.*[11])

There is a caveat to this scenario, however. It is absentee ownership. Let me explain.

Years ago, I worked as a hired hand on cattle ranches in northwestern Colorado. The ranches on which I worked were part of an isolated community in which the ranchers had all been raised in the vicinity of their ranches and had a deep sense of place. Everyone knew everyone else, and a handshake sealed a bargain. There was also a cooperative spirit among the ranchers: from keeping an eye on such things as one another's fields, livestock, and fences to purchasing provisions for neighbors when one of the ranchers, or their wives or hired hands, drove the 80 miles to town.

The ranchers also worked together during summer haying and autumn roundup. There were two tractors among the four ranchers and one old-fashioned hay stacker (no hay bailer in those days). We would start at one person's ranch, put up the hay in what was called "loose stacks," and go on to the next ranch. The person on whose ranch we were haying fed the workers. Whose hay was put up first alternated each year in a sense of fairness.

A few years after I left the ranches, two of them were sold to an eastern corporation of some kind. I went back once and found that the absentee owners and their offspring, who summered every year at the ranches, had no sense of commitment to them, the community, or the surrounding country (=landscape). Consequently, they had used everything to its maximum extractive capacity and had literally run everything into the ground.

Unfortunately, this sort of absentee ownership is the norm, rather than the exception. Absentee owners (especially national and international corporations) are seldom vested in the welfare and sustainability of the community and landscape in which they have economic investments. Their allegiance is to the profit margin and not to the community as a sense of people and place. They are thus willing, more often than not, to break all the rules of sustainability in the name of maximizing short-term monetary gain.

Beyond the Local Landscape

The spatial patterns we see on landscapes, as previously mentioned, result from complex interactions among physical, biological, and social forces. Because most landscapes have also been influenced by the cultural patterns of human use, the resulting landscape is an ever-changing mosaic of unmanaged and managed patches of habitat, which vary in size, shape, and arrangement.

The pattern of changes in the North American forests or the mid-continental Great Plains before Europeans settled there was closely related to topography and to the pattern of Nature's disturbances, especially fire. When the Europeans began to disturb the landscape through such introductions as the grazing of livestock, clearing and plowing the land, and suppressing fire, they altered the patterns of forests and grassland only selectively, because their manipulations were relatively limited to areas adjacent to human settlement. Nevertheless, those human-created disturbances began to cause unforeseen changes in the landscape, changes we are now having difficulty dealing with.

A disturbance is any relatively discrete event that disrupts the structure of a population and/or community of plants and animals or disrupts the ecosystem as a whole and thereby changes the availability of resources and/or restructures the physical environment. We can characterize cycles of ecological disturbances ranging from small grass fires to major hurricanes by their distribution in space and the size of disturbance they make, as well as their frequency, duration, intensity, severity, synergism, and predictability.

In the Pacific Northwest, for example, vast areas of unbroken forest that were at one time in our National Forest System have been fragmented by clearcutting and rendered homogeneous by cutting small patches of old-growth timber, by converting these patches into plantations of genetically selected nursery stock, and by leaving small uncut patches between the clearcuts. Because this "staggered-setting system," as it is called, required an extensive network of roads, almost every water catchment was penetrated by logging roads before half the land area was cut. The whole of the National Forest System thus became an all-of-a-piece patchwork quilt with few, if any, forested areas large enough to support those species of birds and mammals that required the interior of the forest as their habitat.

Changing a formerly diverse landscape into a cookie-cutter sameness has profound implications. The spread of such ecological disturbances of Nature as fires, floods, windstorms, and outbreaks of insects, coupled with such disturbances of human society as urbanization and pollution, are important processes in shaping the landscape. The function of those processes is influenced by the diversity of the existing landscape pattern.

Disturbances vary in character and are often controlled by physical features and patterns of vegetation. The variability of each disturbance, along with the area's history and its particular soil, leads to the existing vegetational mosaic.

The greatest single disturbance to the ecosystem is human disruption—introductions of practices (such as agriculture), substances (such as nuclear waste), and technologies (such as the suppression of fire). These disruptions result most often from our attempts to control the size (minimize the scale) of the various cycles of Nature's disturbance with which the ecosystem has evolved and to which it has become flexibly adapted.

Converting habitats into agricultural fields is the single greatest cause of the extinction of species, which is currently about a thousand times faster than the historical rate, even in geological time. Since 1700, agricultural cropland has increased fivefold worldwide. Some landscapes have been especially hard-hit, according to the United Nations report *Global Biodiversity Assessment*:[62] 98 percent of the dry forest on the Pacific coast of Central America is gone, as are 90 percent of Brazil's Atlantic forest, half of Thailand's mangrove forest, and 54 percent of the wetlands in the United States.

As we struggle to minimize the scale of Nature's disturbances in the ecosystem, we alter the system's ability to resist or to cope with the multitude of invisible stresses to which the system adapts through the existence and dynamics of the very cycles of disturbance that we "control." Today's forest fires, for example, are more intense and more extensive than those in the past because of the build-up of fuels since the onset of fire suppression, which makes them not only increasingly expensive to control but also increasingly dangerous to private property if not controlled.

Many forested areas are primed for catastrophic fire. Outbreaks of plant-damaging insects and diseases spread more rapidly over areas of forest and rangeland that have been stressed through the removal of

Nature's own disturbances, to which they are adapted and which control an area's insects and diseases.

The precise mechanisms by which ecosystems cope with stress vary, but one mechanism is tied closely to the genetic selectivity of its species. Thus, as an ecosystem changes and is influenced by increasing magnitudes of stresses, the replacement of a stress-sensitive species with a functionally similar but more stress-resistant species preserves the ecosystem's overall productivity. Such replacements of species—redundancy—can result only from within the existing pool of biological and genetic diversity. Nature's redundancy must be protected and encouraged.

Human-introduced disturbances, especially fragmentation of habitat, impose stresses with which the ecosystem is ill adapted to cope. Biogeographical studies show that "connectivity" of habitats with the landscape is of prime importance to the persistence of plants and animals in viable numbers in their respective habitats—again, a matter of biological and genetic diversity. In this sense, the landscape must be considered a mosaic of interconnected patches of habitats, like vegetated fencerows, which act as corridors or routes of travel between patches of farm forest, livestock allotments, or other suitable habitats.

Whether populations of plants and animals survive in a particular landscape depends on the rate of local extinctions from a patch of habitat and on the rate with which an organism can move among patches of habitat. Those species living in habitats isolated as a result of fragmentation are therefore less likely to persist. Fragmentation of habitat, the most serious threat to biological diversity, is the primary cause of the present global crisis in the rate of biological extinctions. On public lands much, if not most, of the fragmentation of the habitat is a "side effect" of management policies that stress the short-term production of commodities at the long-term expense of the environment. Actually, however, there are no "side effects"—only unintentional effects!

Modifying the connectivity among patches of habitat strongly influences the abundance of species and their patterns of movement. The size, shape, and diversity of patches also influence the patterns of species abundance, and the shape of a patch may determine the species that can use it as habitat. The interaction between the processes of a species' dispersal and the pattern of a landscape determines the temporal dynamics of its populations. Local populations of organisms that can disperse great distances may not be as strongly affected by the spatial arrangement of patches of habitat as are more sedentary species.

Our responsibility now is to make decisions about patterns across the landscape while considering the consequences of our decisions on the potential cultural capacity of the generations of the future. The decisions are up to us, but one thing is clear: although the current trend toward homogenizing the landscape may help maximize short-term monetary profits, it devastates the long-term biological sustainability and adaptability of the land and thus devastates the long-term sustainability of communities. Greed thus disrupts stability.

It is not the relationship of numbers that confers stability on eco-systems; it is the relationship of pattern. (Similarly, it is the quality, not the quantity, of an economy that determines its relative sustainability.) Stability flows from the patterns of relationship that have evolved among the various species. A stable, culturally oriented system, even a very diverse one, that fails to support these co-evolved relationships has little chance of being sustainable.

To create viable culturally oriented landscapes, we must stop frag-menting landscapes by focusing on such commodity-producing arti-facts as forest clearcuts, agricultural fields, and livestock-grazing allot-ments. Because ecological sustainability and adaptability depend on the connectivity of the landscape, we must ground our culturally designed landscapes within Nature's evolved patterns and take advan-tage of them if we are to have a chance of creating a quality environ-ment that is both pleasing to our cultural senses and ecologically adaptable.

We must move toward connectivity of the landscape. If we are to have adaptable landscapes with desirable cultural capacities to pass to our heirs, we must focus on two primary things: (1) caring for and "managing" for a sustainable connectivity and biological richness be-tween such areas as logging units, agricultural fields, livestock-grazing allotments, and communities within the context of the landscape as a whole and (2) protecting existing biological and genetic diversity, including habitats, at any price for the long-term sustainability of the ecological wholeness and the biological richness of the patterns we create across landscapes, especially in the face of a changing global climate.

Global Climate Change

The average temperature of the surface of the globe is determined by a complex of factors, including the amount of energy received from the sun, the properties on the Earth's surface that absorb the sun's heat (such as the size and distribution of oceans, snowfields, forests, and deserts), and the absorptive properties of the atmosphere. A high proportion of the energy that is absorbed by the lower atmosphere and by the Earth's surface is emitted as heat. Some of the heat passes through the atmosphere into space, but the rest is absorbed into the atmosphere and passed back toward Earth.

At the root of the predicted changes in climate is the accumulation of an array of gases in the atmosphere that trap heat as it is radiated outward from the Earth's surface. Carbon dioxide has to date received the most attention, but other gases are involved as well, including water vapor, ozone, nitrous oxide, chlorofluorocarbons, and methane.

Although there is uncertainty as to how the climate will respond, there is no doubt that greenhouse gases are accumulating. Atmospheric carbon dioxide began to rise in the latter part of the nineteenth century as a result of clearing forests and plowing prairies, both of which released into the atmosphere carbon dioxide that had once been stored in the living tissues of plants and in the soil. Burning of fossil fuels, such as coal, and accelerated deforestation since the 1940s have greatly exacerbated the problem.

Although public attention has focused on the problems caused by deforestation in the tropics, problems also exist in the northern hemisphere. On the one hand, burning of tropical forests converts about 50 percent of the badly needed nitrogen contained in the forests' biomass to a gas, which is lost from the forests to the atmosphere. It also causes vast amounts of smoke, which can alter the internal physical structure of clouds so that severe repercussions may be in store for the water cycle in the tropics. On the other hand, clearcutting old-growth Douglas fir forests in the Pacific Northwest results in a net release into the atmosphere of the carbon dioxide stored in the stems of the live trees. It takes several centuries to recoup this loss, even when the cut forest is replaced with young, fast-growing trees in tree farms.

In many areas of the world, including parts of Canada, the United

States, and Central America, logged forests are not being successfully regenerated with young trees. In Central America, for example, deforestation has been rapidly accelerating during the past three decades. If it is not slowed, there will be little but scrub forest left by the end of this century. If important areas of forest, particularly those that serve the major rivers and water catchments of the region, are not protected soon, no amount of social or economic reform will be able to provide for the many basic needs of the increasing population of the region. This also means that the carbon dioxide lost to the atmosphere by cutting the old-growth forest is not being mitigated by the capacity of young trees to use carbon dioxide and thus remove it from the atmosphere.

With a doubling of the pre-industrial atmospheric concentration of carbon dioxide (which will be reached some time during the next century), temperatures in the conterminous United States are predicted to increase in winter to the equivalent of a four- to six-degree shift southward in latitude. In summer, the temperature increase would be the equivalent of a five- to eleven-degree shift southward in latitude. This would be like shifting southern California to central Oregon, which is a difference of about ten degrees in latitude.

Note that a greater warming is predicted in summer than in winter. The magnitude of warming is also predicted to vary from east to west across the United States and Canada. The greatest warming will take place in the area of the Great Plains, where North Dakota may have a climate similar to that which presently exists in Texas. In addition, the models predict a warming in the mountains that will be equivalent to roughly a 2,000- to 3,000-foot decrease in elevation. This means that most, if not all, of the subalpine and alpine areas would probably disappear in the United States south of the Canadian border.

If current trends in the emission of greenhouse gases continue, increases in carbon dioxide and the other gases are predicted to result in "doubled carbon dioxide" temperatures in less than 50 years. It is therefore the consensus of the mainstream scientific community that there is no alternative to reducing carbon emissions.[63]

Other greenhouse gases, although less abundant in the atmosphere than carbon dioxide, have a much greater warming effect per molecule. For example, one molecule of methane has 3.7 times the potential of one molecule of carbon dioxide for warming the atmosphere.

A single molecule of nitrous oxide has 180 times and chlorofluorocarbon-12 has 10,000 times the potential capacity of one molecule of carbon dioxide to warm the atmosphere.

The models of climatic change are less consistent in predicting the greenhouse effect on precipitation, which is a crucial factor in how ecosystems will respond to changes in climate. However, it is generally agreed that warming will be greater at high latitudes than at lower latitudes. The consequence of such warming will be the narrowing of the global temperature gradient, which in turn seems likely to alter global patterns of precipitation, but what these changes might be remains unclear. Nevertheless, a warming climate, even with no change in precipitation, should increase drought because of the greater evaporative demand brought about by generally higher temperatures.

"We are already feeling the early effects of an altered climate," says Christopher Flavin of the Worldwatch Institute. In northwestern Ontario, Canada, for example, the climatic, hydrologic, and ecological records for the Experimental Lakes Area show that air and lake temperatures have risen by 3.6 degrees Fahrenheit and that the length of ice-free days has increased by three weeks over the last twenty years.[64] (Hydrologic is the adjectival form of hydrology, which refers to the scientific study of the properties, distribution, and effects of water on the Earth's surface.)

Further, higher than "normal" evaporation and lower than average precipitation have occurred, resulting in a decrease in the rate of renewal of water in the lakes. In addition to other changes within the lakes themselves, the concentrations of most chemicals have increased in both the lakes and the streams because of the decreased renewal of the water and because of forest fires in the water catchments. These observations may provide a preview of the effects of increased greenhouse warming on boreal lakes.[64]

In addition, there has been a dramatic warming of Antarctica in recent decades, which has caused a chunk of antarctic glacier the size of the state of Rhode Island to collapse into the South Atlantic. Siberia is now warmer than at any time since the Middle Ages. In northern Europe, there has been a series of warm winters and severe winter storms. Related to this warming trend is the retreat of alpine glaciers, which is exposing ice and rock for the first time in thousands of years. And the interior areas of northern India have experienced life-threatening summer heat waves in recent years.[65]

During a meeting in the late summer of 1995, the Intergovernmental Panel on Climate Change concluded that "a pattern of climatic response to human activities is identifiable in the climatological record." The panel's assessment projects an additional rise of 1.4 to 6.3 degrees Fahrenheit in the average global temperature by 2100. Although the projected increase shows a wide range of variation, even at the lower end it is a faster change than any experienced since human civilization began.[65]

Although this may not seem like much of a change, the average global temperature was only 5.4 to 9 degrees Fahrenheit cooler during the last ice age than it is today. If geological history has anything to teach us, one of its most important lessons may well be that the faster the pace of global warming, the harder it will be for humanity, as well as ecological systems, to adapt to the changes.[65]

It is therefore noteworthy that the Intergovernmental Panel on Climate Change concluded in its 1995 report that any rate of change in temperature above roughly 0.18 degrees Fahrenheit per decade, which is about twice the rate of temperature increase experienced over the last century, could cause considerable havoc. Yet the upper end of the Intergovernmental Panel's projected increase in average temperature (6.3 degrees Fahrenheit) represents a rate of increase that is more than 0.54 degrees Fahrenheit.[65]

It is not so much the relatively modest increase in the projected average global temperature that is of concern to scientists, but rather the possible disruption of atmospheric and oceanic systems that regulate weather. Recent studies indicate that global warming will cause the "extremes" in climate to become more common, which will place an unprecedented stress on ecosystems as well as the human economy. According to the 1995 Intergovernmental Panel on Climate Change, "the incidence of floods, droughts, fires, and heat outbreaks is expected to increase in some regions" as global warming occurs.[65]

In a time when many people live in air-conditioned homes, work in air-conditioned buildings, and eat fresh food grown hundreds or thousands of miles away, it is easy to ignore and then forget about our dependence on the climate. This is especially true for people who do not travel widely. Nevertheless, people still live in areas where the supply of water is adequate, if not abundant, and their nutritional and material requirements are met. Although societies can and do cope with isolated episodes of drought, heat waves, or floods by obtaining relief supplies, such as medicines, water, and food, from elsewhere, simultaneous disruptions of even a moderate magnitude in several regions could be unmanageable.[65]

One possible result of disrupting the global climate is more frequent and prolonged droughts. This could exacerbate an already critical shortage of water that currently plagues 80 countries with 40 percent of the world's population, according to the World Bank. Moreover, the availability of water for agriculture is already a major constraint in many areas of the world and is getting worse as rivers and underground aquifers are gradually being depleted. And this says nothing about urban areas, which are competing for the same water.[65]

Because some of the projected scenarios about global warming show increasing drought in various areas, such as the mid-continental "Great Plains," the greenhouse effect may also alter fire regimes and in some areas contribute to an increased frequency of forest fires. There may be some difficulty, however, in determining potential effects of global warming on the fire regimes, because most historical studies, termed chronosequences, are hampered by effects of unknown historical events, which can result in erroneous interpretations.

It is therefore particularly important to study the major ecological processes in an integrated fashion, because mechanisms are interdependent. From an ecological point of view, the variability in fire regimes is more likely to be important to plant communities, especially forests, than are the mean values computed from some arbitrary period of fire history.

For example, unusually long periods without fire may lead to the establishment of species of plants that are intolerant of fire. The simultaneous occurrence of such fire-free periods and wetter climatic conditions may also be extremely important to such species of plants as ponderosa pine that have episodic patterns of regeneration (specific, discrete episodes) as opposed to plants whose patterns of regeneration form a continual, yearly process.

Therefore, while statistical summaries of fire history over time are useful in understanding the general comparisons of fire regimes in different forests, the influence of fire on the ecosystem is a strongly historical process. Hence, forests of the southwestern United States may be more a product of relatively short-term and unusual periods of climate and fire frequencies than of average or cumulative periods of long-term histories of climate and fire frequencies.

How change in the global climate will affect habitats and the human communities living within them is not known exactly, because we are facing a change in the global climate that is predicted to take place with a speed unprecedented by anything known in geological history.

Climate Change and the Migration of Ecosystems

There have been and again will be drastic changes in habitats that affect whole groups of plants and animals due to changes in climate. Because plants and animals help create a given habitat through their aboveground/belowground symbiotic interactions, in the sense that habitats change, habitats, too, can be thought of as part of the evolutionary process. After all, both plants and animals together help create new habitats as well as the extinction of old ones. To gain a sense of what I mean, consider conditions in eastern North America at the close of the Wisconsin glaciation, about 10,000 years ago.

The modern northern flora and fauna of eastern North America are composed largely of post-Wisconsin glacial-stage plants and animals that immigrated to ground previously stripped of life by glacial ice. Competition therefore favored species adapted to harsh northern environments, species that could disperse rapidly. Groups of animals composed of species from northern and temperate habitats lived on the southern edge of the glacier. Unadaptable temperate species continued to inhabit local areas of relatively unaltered climate, while those that could adapt to some degree survived where the glaciers had not encroached.

As the climate changed, habitats around the glacier slowly changed. Those covered with ice were created and destroyed more rapidly than those along the edge of the ice. The gradual changes created a continuum of small habitats, which supported a richer collection of plants and animals than the flora and fauna in previously glaciated areas.

As the glaciers receded, most mammals followed habitats northward, migrated to higher latitudes, underwent physiological adjustments, or became extinct. The varied habitats and the adaptability of other mammals allowed them to survive by moving southward ahead of the advancing Wisconsin glaciation and then northward again as the glacier melted. Only the less adaptable larger species were particularly prone to extinction.

Habitats change, especially under the influence of a growing human population. Sometimes habitats evolve slowly and gradually and sometimes quickly and drastically, but regardless of the way they do it, all habitats change. When they do, there is a general reshuffling of plants and animals. More adaptable species may for a time survive a change in habitat, even a relatively drastic one, but in the end they too must change, migrate elsewhere, or become extinct.

We humans have changed and are changing the global ecosystem and all of its component habitats at an exponential rate. Today, we have become the major cause of extinctions and of evolutionary leaps. Some ecosystems and their habitats may be able to mitigate the alterations to which we subject them. But alas, most alterations are damaging to the ecosystem as we know it and are prone to spread. Others evolve into ecosystems that we humans find less desirable, often because the new species, which quickly replace those lost, cannot live up to our human expectations.

Today, all habitats pass through the hands of time as well as through the hands of human society, if for no other reason than that we have polluted the atmosphere and, through the air, the soils and the waters of the world.

In addition, the human-caused rate of change in the climate is unprecedented. In mountainous areas, for example, a predicted three-degree rise in temperature would mean that subalpine and alpine habitats, to compensate for the increased temperature, would have to migrate upward in elevation about 1,600 feet. That dramatic a migration would affect the plants and animals of those habitats.

Such migration would lead to a reduction in both the total extent and the number of areas of subalpine and alpine habitats. As a result, those animals that require large home ranges and these specific habitats may become extinct as their habitats shrink. Given this scenario, even a two-degree rise in temperature over the next 50 years could cause the extinction of from 10 to 50 percent of the animals now living in the subalpine and alpine "habitat islands" on the tops of isolated mountains in the Great Basin of the American West.

Further, the major impact of such a fast rate of change will come not from average changes in the weather but rather from striking climatic events, such as prolonged drought in the American Midwest or increased rainfall in the Indian subcontinent. Such climatic changes could lead to increased flooding in both India and Bangladesh.

And speaking of rain, the total amount of rainfall may not be critical for undisturbed habitats, such as forests, but changes in the timing of the patterns of rainfall would be catastrophic for many species, including human communities, that depend on the current weather patterns for part of their life cycles. In addition, increased temperature will cause polar ice caps to melt and sea levels to rise, with potentially devastating effects on habitats and human communities in low-lying coastal areas.

All the evidence of changes in the global climate once again points to the need for habitats and ecosystems to migrate across the landscape, both in elevation and in latitude, as they have done for millennia. Only this time, they will not be able to extend their geographical distributions fast enough to keep up with the pace of climate change now predicted. Thus, a staggering one third of the world's forests could be forced to migrate as a result of the effective doubling of the concentrations of carbon dioxide projected by the year 2100, according to Steven Humburg, a forest ecologist at Brown University.[65] We humans must therefore help the forests by consciously, purposefully keeping corridors of migration free of human-caused obstructions across the landscapes with which we interact as communities, singly and in the aggregate.

As previously stated, the long-term sustainability of human communities depends on the long-term sustainability of the habitats and ecosystems of which they are a part. In the face of rapid change in the global climate, sustainability means protecting the ability of habitats and ecosystems, and human society along with them, to adapt to change by wandering at will across landscapes, which brings up the critical issue of water.

WATER

Water is a physical necessity of life. Water is perhaps the most important commodity when it comes to the sustainability of a community. A community's supply of quality water is therefore precious almost beyond compare.

The amount and quality of water available for human use are largely the result of climate and strategies for taking care of the ecological health of water catchments. In North America, sustaining the health of water catchments is particularly important in order to protect the annual snowpack from which the vast majority of all usable water comes. However, protecting the quality and quantity of society's water supply is not a primary consideration of timber corporations, which operate where most of the snow falls.

People seldom realize that drinkable water comes predominantly from forested water catchments. Even much of the prehistoric ground water that is pumped to the surface for use in agriculture came from forested water catchments. Water, and therefore hydroelectric power, is a forest product just as surely as is wood fiber.

A curious thing happens, however, when water flows outside the forest boundary: we forget where it came from. We fight over who has the "right" to the last drop, but pay little attention to the supply—the health of the forested water catchments.

As a nation with once bountiful resources, the United States has rarely faced limits to those natural resources. However, present trends and experience indicate that every additional drop of water conserved and thus available enables more economic growth, which further raises the demand for more water and more economic growth. Effective management of water will thus necessitate attention to both demand and supply.

The availability of water for agricultural use varies by location and over time. Availability of water also depends on such variations in components of the hydrologic cycle as precipitation, evaporation, transpiration, infiltration, and runoff. Because these components are interrelated, a change produced by technology in one component of the cycle will inevitably affect other components.

In the short history of the United States, there have always been more lands and more resources to exploit and a philosophy that technology could supplement natural resources when needed. Today, however, stretching such water resources to accommodate the continuing economic growth of the western United States while protecting existing patterns of water use will require levels of technical development that are increasingly damaging ecologically and no longer feasible economically. Most people do not realize that only a small part of the water used in the United States goes to towns and cities. The overwhelming share is used for irrigation.[66–69]

For example, withdrawals of water for irrigation range from 80 percent of the total use in Utah to 90 percent in New Mexico. Further, the use of water for irrigation is inefficient at best, as shown by a U.S. Geological Survey, which found that the loss of water by seepage from canals was one third of the amount actually delivered to irrigated farms. And this does not include the loss of water to the atmosphere through evaporation.[66-69]

According to Professor Luna B. Leopold, the persistence of the pro-economic expansion bias of the U.S. Bureau of Reclamation is increasingly inexcusable. This attitude is still held in spite of the obvious strain on both the quality and the quantity of the supply of water. He says, "It is deplorable that the government agency most responsible for managing water in water-short regions continues to be so insensitive to the hydrological continuum and the equity among claimants."[66]

The hydrological continuum, as used by Leopold, is different from the hydrological cycle. The hydrological cycle continues for better or for worse, but the idea of a hydrological continuum implies the maintenance of a quasi-equilibrium operational balance among the processes within the hydrological cycle, which involve the air, water, soil, biosphere, and people. In other words, if withdrawals of water are balanced with Nature's capacity to replenish that which is used, the use of water can be measured in such a way that the available long-term supply is protected.[66-69]

There are thus two options in managing the use of water. One is to protect the availability of the long-term supply by disciplining ourselves to use only what is necessary in the most prudent manner. The other is to take water for granted and use all we want with no discipline whatsoever (as we do now through continual economic expansion) and then wonder what to do when faced with a self-inflicted shortage.

By using all the water we want in a totally undisciplined manner, we are insensitive to both the care we take of the water catchments in each bioregion and the speed with which we mine the supply of stored available water. As stated by Professor D.J. Chasan, "One might suppose that people would automatically conserve the only naturally occurring water in a virtual desert, but one would be wrong. Land and farm machinery have capital value. Water in the ground, like salmon in the sea, does not. Just as salmon are worth money only if you catch them, water is worth money only if you pump it."[70] We are therefore pumping groundwater, and we are damming, diverting, and channelizing the rivers to "tame" and "harness" their water for short-term use based on poor economics, rather than nurturing the environment to ensure the availability of an adequate long-term supply of water.

For instance, the U.S. Army Corps of Engineers designs and builds structures to control flooding and to improve navigation, but it also issues permits for the alteration of bodies of water, marshlands, and estuaries. The Soil Conservation Service, which was originally a land management agency, became another engineering organization following World War II. Where earlier policy had been in the hands of agronomists, soil technicians, and managers of rangelands, following the war most of the policymaking positions were filled by engineers.[66]

One aspect of the engineering programs of both the U.S. Army Corps of Engineers and the Soil Conservation Service is the practice of "channel improvement" in streams and rivers. Such "improvements" lead to a straightening of a stream's channel and a change of its shape, which in turn destabilizes the channel. Destabilization of the channel causes downstream effects, such as erosion of the banks, alterations of the channel's bed, degradation of the aesthetics, and changes, often considered to be undesirable, in the composition of the plants and animals that inhabit the stream. From these numerous "improvements," each planned on its own isolated rationale, comes the next larger-order magnitude of massive flooding,[66] such as occurred recently in the lower reaches of the Mississippi River.

No studies, according to Leopold, have been conducted by either the Corps of Engineers or the Soil Conservation Service to determine the long-term effects, either on-site or downstream, of such alterations in the channel. The interest of these government agencies, in the West at least, has not been in the long-term future of the landscape because the computed, but often unrealistic, cost–benefit ratio is on the side of utility.

Is water to become the ultimate economic/political/environmental club with which we bludgeon each other? This question is appropriate here because as we witness the demise of the old-growth forests, we are also limiting the available supplies of potable water. The only solution is an environmental one: protecting the health of water catchments on a landscape scale that first and foremost nurtures the health of soil and water, lest everything else become unhealthy. Like migratory birds and anadromous fish, environmental crises, such as the pollution of air and water, know no political boundaries. Soil, water, air, and climate form a seamless whole, the thin envelope we call the biosphere, which is all we have in all the Universe.

With the growing realization of the ecological interdependency among all living forms and their physical environment, it can hardly be doubted that even "renewable" resources show signs of suffering from the effects of society's unrelenting materialistic demands for more and more. These demands have degraded the renewability of resources in both quality and quantity. Water can be thus characterized, because it is increasingly degraded by soil erosion, increases in temperature, and pollution with chemical wastes, salts from irrigation, and overloads of organic materials. Is it any wonder, therefore, that the hydrological system is under stress?

The rub lies in the available technology. Many farmers, interested only in the short-term production of their own fields, are still plowing uphill and downhill, despite 50 years of soil conservation. The soil eroding from their fields, which is changing the conditions downstream, is not their problem. Similarly, the county agent who advises the farmers is often more likely to be concerned with the farmers' fields than with the health of the whole river basin. Because most water supply engineers see the hydrological system as a whole as being outside of their domain, they are not immediately concerned with its problems. We therefore rely on our technology to provide safe, chemical-laden water, but we are managing water after the fact.

As with any problem, there are solutions, but we tend to look for solutions only where the symptoms are obvious. For example, I used to live in an old mobile home. You know the kind: it leaked every time a dark cloud appeared on the horizon. Finding where the leak ended, over my desk or over my pillow, was no problem, but finding the source of the leak was often so difficult that I had to repair the entire waterproof coating on the roof—and then hope for the best.

How We Think About Water Catchments

The problem with water catchments begins with the headwaters, the first-order stream and its catchment basin. A first-order water catchment is always a special case; in fact, it is probably the only part of the land that is manipulated, where the hydrology has ecological integrity. Further, it is the headwaters and therefore controls the initial quality of the water for the whole drainage basin.

A first-order water catchment, by definition, is unique. A second-order water catchment is unique among second-order water catchments but is a common denominator, an integrator, of the first-order water catchments that created it. A third-order water catchment is unique among third-order water catchments but is a common denominator of the first- and second-order water catchments that created it, and so on.

Our thinking, and therefore our view of the world, is generally limited to a kaleidoscope of special cases because we choose to focus on "discrete" parcels of real estate. If we deal only with special cases (a mile of stream, for example), we perpetuate our inability to understand that particular mile of stream, the entire stream, and the water catchment as a whole. If, on the other hand, we deal with a particular mile of stream (a special case) in relation to the whole water catchment (the common denominator), we enhance our ability to understand both the mile of stream and the water catchment because each is defined by its relation to the other. Understanding how a reach of stream relates to the whole water catchment is like understanding how a single chair relates to a room.

If you stand in the doorway and survey the room, you will see the chair both in the room and in relation to the room, but when you focus only on the chair you can no longer see the room or the chair's relationship to it. Unfortunately, most people in a community do not see that first-order water catchments are the initial controllers of water quality for supplies of domestic water.

We therefore cut timber down into the stream bottoms of both first- and second-order streams, even in municipal water catchments, because the timber is thought to have greater immediate economic value than the water. Moreover, because politically important fish, such as salmon and steelhead trout, do not live at the high elevations in which most of these small streams occur, the water is deemed to be of no visible economic importance. The invisible importance of the water in a water catchment, far from the tap that dispenses it, becomes visible when the water reaches human communities and becomes usable. But first, it must be stored somewhere to be available when needed.

The Storage of Water

The storage of a community's water originates far from the tap you turn to fill a glass with that most precious of liquids. Water is stored in four ways: (1) in the form of snowpack aboveground; (2) in the form of water penetrating deep into the soil, where it flows slowly belowground; (3) in belowground aquifers and lakes; and (4) in aboveground reservoirs.

Most water used by communities comes first in the form of snow, either at high elevations or northern latitudes, which subsequently feeds the streams and rivers that eventually reach distant communities and cities—rivers such as the Columbia, Snake, Colorado, Missouri, Mississippi, and so on. Snowpack is aboveground storage, which, under good conditions, can last as snowbanks late into the summer.

How much water the annual snowpack has and how long the snowpack lasts depends on five things: (1) the timing, duration, and persistence of the snowfall in any given year; (2) how much snow accumulates during a given winter; (3) the moisture content of the snow (wet snow holds more moisture than dry snow); (4) when the snow begins melting and the speed at which it melts (the later in the year it begins melting and the slower it melts, the longer into the summer its moisture is stored aboveground); and (5) the health of the water catchment.

Although the first four points seem self-evident, the last one requires some explanation. In dealing with the health of water catchments, one must consider those of both high and low elevation.

How we treat our high-elevation forests (and those at more northerly latitudes) is how we treat a major portion of the most important source of our supply of potable water. If high-elevation forests are clearcut, as is now happening at an exponential rate, then the snow melts and runs off early (most of it flowing over the surface of the ground), which usually overflows low-elevation reservoirs and is lost for use in late summer and autumn. If, on the other hand, high-elevation forests are healthy, snow melts more slowly and the would-be runoff can infiltrate deep into the soil, moving downslope in slow-motion storage, to be available in streams, rivers, and reservoirs during late summer and autumn, when it is most needed.

Thus, when considering the supply of water for communities, there are forested areas, particularly at high elevation, that humility, wisdom, and long-term economics dictate should not be cut even once, regardless of the perceived immediate, short-term dollar value of the wood fiber. To protect such areas for the storage water in the form of snowpack will require a drastic shift in thinking, because at present the only economic value seen in high-elevation forests is the immediate extraction of wood fiber. Nevertheless, the treatment received by high-elevation forests, which catch and store water, affects all human communities, from the smallest rural village to the largest city.

In contrast, most low-elevation water catchments must be much larger in area than a high-elevation catchment to collect and store the same amount of water. Although snow may not be as important for the storage of water in low elevations, the ability of water to infiltrate deep into the soil is equally important. The storage of water at low-elevation, nonforested areas is often in wetlands, subterranean aquifers and lakes, as well as in aboveground reservoirs. Regardless of where the water catchment is, however, roads have a tremendous effect on both the quality and quantity of water that ultimately reaches a community.

Roads and Water

In addition to humanity's lack of vision with respect to nurturing water catchments for the retention of water, roads therein affect the quality, quantity, and distribution of water in the soil of the catchment, regardless of whether they are graveled and constructed to extract timber or paved and constructed as access to homes in a housing development. Roads bleed water from the soil the same way cuts in the bark bleed latex from a rubber tree or sap from a sugar maple.

The construction and use of a road severely disturb the soil, which increases the rate of runoff, reduces the flow of subsurface water, and alters the equilibrium of shallow groundwater. Unfortunately, the information needed to understand the effects of a road on the regime of surface and subsurface water is limited.[71]

Unless water infiltrates deep into the soil of a water catchment, it runs downhill and reaches the cut bank of a logging road or even a major highway, which brings it to the surface, collects it into a ditch, and puts it through a culvert to begin infiltrating again. The water then meets another road cut, and so on. Water is sometimes brought to the surface three, four, or more times before reaching a stream. Water is purified by its journey through the deeper soil, but not by flowing over the surface of the ground.

In fact, ditches and gullies, such as those that form on the downhill side of culverts that pass under roads, function effectively as pathways down which water flows. The denser the network of roads, the greater the drainage of water over the soil's surface, and the less time it takes for peak flows to occur.[71]

But how deep is deep enough to avoid the ditches at the bases of banks alongside roads? I have seen roadbeds blasted out of solid rock to depths of 50 or 60 feet, and I have seen water seeping out of this same rock into the roadside ditches in July and August, which is symptomatic of the disruption in the flow of water. This means bringing precious water to the surface of the ground, where it not only evaporates but also becomes polluted by sediment, oil, and chemicals from the road and human garbage in the ditch. Roads therefore have an impact on the hydrological cycle of a water catchment and on the purity of the water that ultimately reaches human communities.

Disrupting the flow of water through the soil on steep slopes, even forested slopes, can cause instability and increase erosion during a severe rainstorm or as snow melts. Such conditions in the vicinity of seeping water will cause soils to become saturated with little or no infiltration, which in turn weakens them and leads to greater local runoff of water over the surface of the soil and hence greater erosion.[71]

In housing developments within a water catchment surrounding a community, on the other hand, roads and streets are paved, which creates an impervious coating over the surface of the land. This impervious layer prevents the water, both rain and melting snow, from infiltrating into the soil, where it can be stored, further purified, and recharge existing wells. The water instead remains on the surface of the roads and streets, where it mixes with pollutants that collect on the pavement.

Because paved roads and streets are lined with curbs and gutters, the now-polluted water is channeled from the paved surface into a storm drain. In addition, each house has an impervious roof, which collects water and channels it into gutters along the edge of the roof. Upon collecting water, the gutters channel it, more often than not, out to the street, where it joins water from the street going down the storm drain. It is then conducted either directly into a sewage treatment plant or directly into a ditch, stream, or river.

In any event, the water is not usable by the local people. Beyond that, the storm water either adds to the cost of running the treatment plant, where it must be detoxified, or it pollutes the waterways from the house roof or pavement of its origin into the ocean, as previously discussed.

The effect of roads and the area covered by houses, both of which eliminate the infiltration of water, is cumulative. Enough roads over time can alter the soil–water cycle as it affects a given community. Remember that the quality and quantity of water is an ecological variable, but most economists and "land developers" consider it to be an economic constant.

Even if water were a constant, a variable is introduced with construction of a single road. The variable is compounded by constructing and maintaining multiple logging roads to extract timber. In addition, logging and plantation management alter the water regime, which affects how the forest grows. Thus, a self-reinforcing feedback loop of ecological degradation in a water catchment is created, altering the soil–water regime, which in turn alters the sustainability of the forest, which in turn affects the soil–water regime, and so on. Eventually, the negative effects are felt in those communities that are dependent on a given water catchment or drainage basin for their supplies of potable water.

One of the main problems is that resource management agencies (federal, state, county, and city, especially those responsible for water) lack a long-term perspective and suffer from a shortage of cooperation and coordination among a truly public-minded leadership. Such agencies are not guided by an ethos of long-term sustainability. Instead, they all too often engage in battles over turf, while doing their utmost to avoid both responsibility and accountability for their behavior. They are guided by a plague of special interests and a disdain for equity. As a result, local communities are the continual losers.

We can continue to degrade and impoverish our supply of water, or we can risk abandoning our conventional thought pattern and, with a strong, concerted commitment, reverse the trend. In the final analysis, we must remember that only so much water is available, and with a change in the global climate, that amount may become even more variable and unpredictable than it already is. And more water cannot be found in the courtroom, no matter how hard we try or who holds the priority rights to the water already available. It behooves us, therefore, to consider how we care for the sustainability of water catchments—lest the wells go dry.

From Wells to the River

The purpose of the preceding scenario is to show that wells can go dry. When wells do go dry, as they are now doing around the outskirts of my home town, there are four possible explanations that I can think of: (1) wells have been overdrawn, (2) drought has depleted the stored available water and there has not been enough precipitation in any form to replenish it, (3) the health of the water catchments that supply the water is sufficiently degraded to limit the supply, and (4) all of the above. This could change with warming of the global climate, however, in which case it is conceivable that number two above would take over and the supply of available water would simply diminish.

In my home community, for example, the wells have not been overdrawn and we have not witnessed a drought in the last couple of years. But the water catchments here in the Pacific Northwest, including that of the major river from which my town gets most of its water, have been largely stripped of forest and laced with roads. They are in terrible shape ecologically. I know because I have flown over them within the last two years. Be that as it may, what is important here is what happens when the wells go dry.

First, those people who have their own wells are not using water supplied by the community, which means that the capacity of the community's supply of water is figured without considering those people who have wells. Thus, when the community draws an allocated amount of water from the river, under the auspices of its "water rights," those people with their own wells are excluded. When the community determines it must increase the size of its water purification plant, the owners of wells are once again omitted from the calculations that determine the volume to be treated for future use.

But what happens when the wells go dry? The people who have wells must have water to survive, and there is no substitute. At some point, they must be served with water from the community. This means that, without increasing the human population, the community must use more of its available water, which means less available overall in case of a protracted shortage.

Is this really a problem? You might think that not *that* many people use wells, and in any case, it's a big river. Both assumptions may be true, but then, how many people does "that" refer to? And remember, I am only talking about one community.

Now, if the community happens to be the first, or even the second or third, to have water rights along the upper reaches of the river, that is one thing. But if the community is the twentieth or thirtieth to have water rights along the same river, or its tributaries, that is something else altogether.

If wells are going dry in one community because of an unhealthy water catchment, or perhaps even a whole unhealthy drainage basin, they are going to go dry in other communities that share the same catchment or basin, which greatly compounds the problem the farther downriver one goes. This poses a difficult question: How does one justly adjudicate water rights? Is it based on first come, first serve? Is it contingent on a community's location along the river continuum? Is it founded on the number of people within a given community? Is it determined by the ratio of agricultural use to household use?

Regardless of how it is done, the river has only so much water. While it may be possible to increase the supply by healing the water catchments within the drainage basin, that will take years. How can the growing population along the entire river be accommodated in the interim? What happens if global warming decreases the overall supply of water by lowering the annual amount of total precipitation?

Should any of these things happen, it may be necessary to eliminate some agricultural use of the water in favor of household use, which will affect supplies of locally grown food, not a bright prospect for a community striving for sustainability and greater economic independence. Economic growth would be drastically curtailed, and the value of private property, both agricultural and urban, would plummet.

Yet, with all these negative possibilities, federal, state, county, city, and rural community governments often refuse to deal in any coherent, cooperative, and coordinated way with the health of water catchments and drainage basins on which municipalities depend for potable water. I have tried for years to get these ideas across, but almost every official with whom I have spoken politely shrugged his or her shoulders, looked appropriately helpless, and promptly passed the buck. Those few—those very few—who understood what I was saying were not in a position of sufficient authority to act, other than through the political chain of command, where they met with the same helpless shrugs that I did.

For example, I sat for almost three years on an environmental advisory committee for my local area. During that time, the decision was made to increase the capacity of the community's water purification plant to accommodate building another 5,000 family homes. I told the officials that it would be wise to limit the growth of the population because if they did not, the community would one day run out of available water. I suggested that they build only 3,000 homes and hold in reserve the water for the other 2,000 for the inevitable shortages during dry years and other unforeseen emergencies.

I was told, however, that they could neither limit the population nor slow economic growth; when the supply of water began to get low, they would figure out some way to increase it. What they were really saying was that *someone else* would have to deal with the crisis, so they opted for the easy way out—passing the buck to some generation in the future.

When available water becomes a limiting factor in sufficient degree to cast serious doubt on the future of its supply, the value of real estate will dry up with the water. People upriver, who become increasingly concerned about their own survival and the value of their own property, also become, in my experience, increasingly self-centered out of fear of loss and do whatever they can to save themselves, even at the expense of those downriver. In the end, because a few people in positions of leadership, lacking moral courage and political will, refused to discipline themselves and act while there was time, everyone—the present generation and those of the future—will lose, because there is no substitute for water.

As American author Harriet Beecher Stowe said: "Private opinion is weak, but public opinion is almost omnipotent." What will it take to help people stuck in the current self-serving world view of an ever-expanding economy to see that the limitations of natural resources are real? What will it take to help them see that technocratic/political fixes will no longer work, that fundamental change is necessary?

There is still time to resolve the problem of water, but it will require moral courage, self-discipline, strong pressure from those communities working toward social/environmental sustainability, and political will (=real leadership) on the part of people in positions of leadership. In addition, it will require reforming the political process that allows corporations and large special-interest donors to contribute to political campaigns in such a way that they "buy" politicians. And it will require open space.

OPEN SPACE

Open space, like water, is available in a fixed amount. Unlike water, however, open space is visibly disappearing at an exponential rate. Once gone, it's gone—unless, of course, rural communities, and perhaps even cities, are torn down to reclaim it. The ability and commitment to maintain a matrix of open spaces within and surrounding a community is critical to the sustainability and ultimately the economic viability of the community, especially a small community in a nonurban setting. Although there are multiple reasons why a community might want to save open spaces, the protection of local water catchments is a crucial one.

Water

Water, as previously mentioned, is a nonsubstitutable requirement of life and is finite in supply. Its availability throughout the year will determine both the quality of life in a community and consequently the value of real estate. It behooves a community, therefore, to take any possible measure to maximize and stabilize both the quality and quantity of its *local* supply of water.

By local supply, I mean water catchments in the local area under local control, as opposed to water catchments in the local area under the control of an absentee owner with no vested interest in the community's supply of water. Such absentee ownership could be a person, corporation, government body, or agency beyond local jurisdiction.

With this in mind, it is wise to purchase as much of the local water catchments as possible and maintain them as open space expressly for the purpose of storing water in the ground, where it can purify itself as it flows slowly toward the wells it recharges. This will help prevent those people with wells from needing municipal water, which will help to maintain a more predictable demand over time. And those people with wells, who do not pay for municipal water, could be charged a fee for using water from the community-owned water catchment as a means of helping defray the costs of maintaining the catchment's health.

If outright purchase of a water catchment is not possible, a community could conceivably enter into a long-term lease or contract to rent the catchment, with control over what is done on it. Then it might be possible to accrue monthly or annual payments toward the price of purchasing the land at a later date. Such an arrangement could benefit the owner in terms of a steady income at reasonable tax rates, while allowing some acceptable use of the land.

Another alternative might be a tax credit payable to the land owner if the community could work in conjunction with the owner to protect the water catchment's inherent value to the community itself. There are probably other options, but the important consideration is to secure the purchase of local water catchments in community ownership as part of the open space program to maintain and protect the quality of life and the local value of real estate. An added value may be that some part of a water catchment could also be used for recreational space for the community as a whole.

Communal Space

Open space for communal use is not only central to the notion of community but also is increasingly becoming a premium of a community's continued livability and the stability of the value of its real estate. Of course, continual economic growth, at the expense of open space, will line the pockets of a few people in the present, but it will ultimately pick the pockets of everyone in the future.

For communal open space to have maximum value over time, however, the community must have a clear vision of what it wants so that the following questions can be answered in a responsible and accountable way: (1) What parcels of land are wanted for the communal system of open space? (2) Why are they wanted? What is their functional value: capture and storage of water, habitat for native plants and animals, local educational opportunities, recreation, aesthetics? (3) How much land is necessary to fulfill numbers one and two? (4) Can one project the value added to the quality of life and/or the consequential value of real estate in the future, including that outside the community's urban growth boundary?

Surrounding Landscape

The land surrounding a community's municipal limits gives the community its contextual setting, its ambiance, if you will. The wise acquisition of open spaces in the various components of the surrounding landscape, whether Nature's ecosystem or culture's, protects, to some extent at least, the uniqueness of the community's setting and hence the uniqueness of the community itself. And the value added, both spiritual and economic, will accrue as the years pass.

Agricultural Cropland

According to C.J. De Loach, "...the objective of agriculture is to encourage the growth of a foreign organism, a crop, at a high density and to suppress...organisms that might compete with it...."[72] Yet it was not always so cut and dried, as noted by David Pimentel: "When man dug holes here and there and planted a few seeds for his food, ample diversity of species remained, but this resulted in small crop yields both because of competition from other plants (weeds) and because insects, birds, and mammals all took their share of the crop."[73]

In modern agricultural practice in North America, however, large fields are often planted with a single species. This specialization has resulted from an ever-expanding centralized corporate power base, aided by technology in an increasingly mechanized society.

In the process of centralizing corporate power, a greatly simplified, and therefore increasingly fragile and labor/energy intensive, environment has been created through the following changes in the land:

1. Increased specialization of farms (growing fewer crops in larger fields) caused amalgamation of small, individual fields.
2. Increased size of individual farms due to specialized corporate farms replacing small, diversified family farms.
3. Increased use of modern machinery that is more easily and more economically operated in large single-species fields.
4. Increased clearing of fencerows to gain more land for agriculture, where one mile of fencerow may occupy one half acre.[74]
5. Increased use of large sprinkler irrigation systems that eliminate uncultivated irrigation ditches and their banks.
6. Replacement of many uncultivated earthen banks of irrigation ditches with concrete.
7. Constant human control of crops with fungicides, herbicides, insecticides, and rodenticides, or all four, if the desired production is to be forthcoming.

8. Federal aid to farmers through the Agricultural Stabilization and Conservation Service for various types of land "reclamation."

As these factors reduced the habitat for many species of wild plants and animals, they also increased the tendency for these same plants and animals, which now surround the croplands, to be perceived as exerting a constant negative influence on production. When wild species, especially animals, use agricultural crops as habitat, they are normally termed "pests." However, whether or not a species is a pest is a matter of perception based on some level of competitive tolerance, which wanes rapidly when money is concerned.

Small, diversified family farms were excellent habitat for wildlife. They provided increased structural diversity and therefore increased habitat diversity through a good mix of food, cover, water, and mini open spaces within surrounding, otherwise rather homogeneous, croplands.

The many small, irregular fields with a variety of crops created an abundance of structurally diverse edges, and tillage offered a variety of soil textures for burrowing animals. Uncultivated fencerows and ditch banks provided strips that not only acted as primary habitat for species, such as insectivorous songbirds, but also provided travel lanes between fields for other species.

Replacement of small family farms by large ones, dependent on mechanization and specialized monocultural crops, caused a drastic decline in wildlife habitats within and adjacent to croplands. Because of the decreased crop stability—increased crop vulnerability—resulting from the greatly simplified "agricultural ecosystem," farmers are more and more inclined to view wild or nonagricultural plants and animals as actual or potential "pests" to their crops.

In addition to stripping habitats from fencerows surrounding fields to maximize tillable soil and get rid of unwanted plants and animals, modern agriculture is killing the soil and poisoning the water with chemicals, which clearly is neither biologically nor culturally sustainable. How can such destructive agriculture be redeemed?

It can be redeemed by meeting with the local farmers and discussing the kinds of produce that could be grown to make the community as self-sufficient as possible. The economic viability of the remaining small, family farmers can be ensured by loyally purchasing their produce. Organically grown produce may cost a little more, but it is healthier, and organic farming heals the soil and does not add polluting chemicals to the water.

A community could purchase open space in the form of fencerows along which to allow fencerow habitat to recreate itself. Then, in addition to mini habitats in and of themselves, the few uncultivated yards could once again act as longitudinal corridors for the passage of wildlife from one area to another. Living fencerows would also make the landscape more interesting, more appealing to the human eye, and add once again the songs of birds and the colors of flowers and leaves to the passing seasons.

The point is to find out what worked sustainably in the past and begin recreating it in the present, and where problems arise, to work together to resolve them. The only way to create, maintain, and pass forward the sense of community is by working together, because the friendliness of a community is founded on the quality of its interpersonal relationships, of which small family farmers are an integral part.

Forest Land

If a community is in a forest setting, the forest more likely than not is a major contributor to the community's image of itself, in addition to which it may comprise an important water catchment. Furthermore, if the community is, or has been, a "timber town," then most of the forest may well have been converted to economic tree farms; therefore, maintaining an area of native forest may be of even greater value. And if some old-growth trees are included in the area, its spiritual value may well be heightened and its value as habitat for some plants and animals greatly enhanced.

On the other hand, if what surrounds a community is no longer forest but rather an economic tree farm, a purchased area could be allowed to evolve once again towards a forest. As such, its aesthetic and spiritual values will increase, as will its potential educational value. Much can be learned by comparing a relatively sterile tree farm with a real forest.[75,76] One will find, for instance, that a forest harbors a far greater diversity of species of both plants and animals than does a tree farm, even one near the age of cutting.

I have used the forest as an example only because I grew up in one, but the same concepts can be applied anywhere. Outside of

Denver, Colorado, for example, is a wonderful open space, which represents a vestige of native short-grass prairie, which once covered that part of the state. It is beautiful! And it is inspirational, creating, as it does, a tangible tie to a now intangible past and an unknowable future.

Riparian Areas and Floodplains

Riparian Areas

Riparian areas can be identified by the presence of vegetation that requires free or unbound water and conditions more moist than normal. These areas may vary considerably in size and the complexity of their vegetative cover because of the many combinations that can be created between the source of water and the physical characteristics of the site. Such characteristics include gradient, aspect of slope, topography, soil, type of stream bottom, quantity and quality of the water, elevation, and the kind of plant community.

Riparian areas have the following things in common: (1) they create well-defined habitats within much drier surrounding areas, (2) they make up a minor portion of the overall area, (3) they are generally more productive than the remainder of the area in terms of the biomass of plants and animals, (4) wildlife use riparian areas disproportionately more than any other type of habitat, and (5) they are a critical source of diversity within an ecosystem.

There are many reasons why riparian areas are so important to wildlife, but not all can be attributed to every area. Each combination of the source of water and the attributes of the site must be considered separately. Some of these reasons are as follows:

1. The presence of water lends importance to the area because habitat for wildlife is composed of food, cover, water, and space. Riparian areas offer one of these critical components, and often all four.
2. The greater availability of water to plants, frequently in combination with deeper soils, increases the production of plant biomass and provides a suitable site for plants that are limited elsewhere by inadequate water. The combination of these factors leads to increased diversity in the species of plants and in the structural and functional diversity of the biotic community.

3. The dramatic contrast between the complex of plants in the riparian area with that of the general surrounding vegetation of the upland forest or grassland adds to the structural diversity of the area. For example, the bank of a stream that is lined with deciduous shrubs and trees provides an edge of stark contrast when surrounded by coniferous forest or grassland. Moreover, a riparian area dominated by deciduous vegetation provides one kind of habitat in the summer when in full leaf and another type of habitat in the winter following leaf fall.

4. The shape of many riparian areas, particularly the linear nature of streams and rivers, maximizes the development of edge effect, which is so productive in terms of wildlife.

5. Riparian areas, especially those in coniferous forests, frequently produce more edges within a small area than would otherwise be expected based solely on the structure of the plant communities. In addition, many strata of vegetation are exposed simultaneously in stairstep fashion. This stairstepping of vegetation of contrasting form (deciduous versus coniferous, or otherwise evergreen, shrubs and trees) provides diverse opportunities for feeding and nesting, especially for birds and bats.

6. The microclimate in riparian areas is different from that of the surrounding area because of increased humidity, a higher rate of transpiration (loss of water) from the vegetation, more shade, and increased movement in the air. Some species of animals are particularly attracted to this microclimate.

7. Riparian areas along intermittent and permanent streams and rivers provide routes of migration for wildlife, such as birds, bats, deer, and elk. Deer and elk frequently use these areas as corridors of travel between high-elevation summer ranges and low-elevation winter ranges.

8. Riparian areas, particularly along streams and rivers, may serve as forested connectors between forested habitats or elevational habitats, such as grasslands. Wildlife may use such riparian areas for cover while traveling across otherwise open areas. Some species, especially birds and small mammals, may use such routes in dispersal from the original habitats. This may be caused by the pressures of overpopulation or by shortages of food, cover, or water. Riparian areas provide cover and often provide food and water during such movements.

In addition, riparian areas supply organic material in the form of leaves and twigs, which become an important component of the aquatic food chain. Riparian areas also supply large woody debris in the form of fallen trees, which form a critical part of the land/water interface, the stability of banks along streams and rivers, and instream habitat for a complex of aquatic plants as well as aquatic invertebrate and vertebrate organisms.[77]

Setting aside riparian areas as undeveloped open space means saving the most diverse, and often the most heavily used, habitat for wildlife in proximity to a community. Riparian areas are also an important source of large woody debris for the stream or river whose banks they protect from erosion.[77] Furthermore, riparian areas are periodically flooded in winter, which, along with floodplains, is how a stream or river dissipates part of its energy. It is important that streams and rivers be allowed to dissipate their energy; otherwise, floodwaters would cause considerably more damage than they already do in settled areas.

Floodplains

A floodplain is a plain that borders a stream or river that is subject to flooding. Like riparian areas, floodplains are critical to maintain as open areas because, as the name implies, they frequently flood. These are areas where storm-swollen streams and rivers spread out, decentralizing the velocity of their flow by encountering friction caused by the increased surface area of their temporary bottoms, both of which dissipate much of the floodwater's energy.

It is wise to include floodplains within the matrix of open spaces for several other reasons: (1) they will inevitably flood, which puts any human development at risk, regardless of efforts to steal the floodplain from the stream or river for human use (witness the Mississippi River); (2) they are critical winter habitat for fish;[77] (3) they form important habitat in spring, summer, and autumn for a number of invertebrate and vertebrate wildlife that frequent the water's edge;[77] and (4) they can have important recreational value.

Try to steal land from a stream or river, and sooner or later it will reclaim it, at least temporarily, and at great cost to the thieves, which brings us to the economics of sustainability.

THE ECONOMICS OF SUSTAINABILITY

It is my intent in this section to point out that the current economics of the global economy are neither ecologically sustainable nor socially beneficial to most people. It is not my intent to analyze why this is the case. For those who might question why I am writing about economics at all, I point out that it does not take an economist to recognize that such things as global deforestation, overfishing, and desertification of land lead again and again to the environmental/social arrogance of many major corporations.[78]

Some corporations, for example, move out of the United States, where the cost of labor is high and expensive equipment is required to control pollution, both of which cut into the corporate profit margin. They move to countries like Mexico, where labor is cheap, they are free to pollute the environment with impunity, and corporate profits can be maximized, regardless of the environmental costs for everyone else in the world, present and future. Other corporations physically destroy the environment in grossly intrusive ways, such as strip mining minerals or clearcutting and burning tropical forests for both the coveted tropical woods and to produce pastures for beef cattle. They ignore all the scientific and social data, which point out that such behavior is immediately devastating to the local cultures and ecologically untenable for future generations.[78]

Such data make no difference, however, because the corporations are simply intent on controlling the economy of global politics and the politics of the global economy, often by financing such political tyrants as purchasable puppets and dictators. And in the process, they are destroying the democratic system of government.[79] I have seen the ecological/social malfeasance of corporations from the industrialized nations, particularly the United States, in every country I have worked in. And it all revolves around profit and the power to control.

The Dynamics of Scarcity

According to a song popular some years ago, "freedom's just another word for nothing left to lose," which in a peculiar way speaks of an apparent human truth. When one is unconscious of a material value, one is free of its psychological grip. However, the instant one perceives a material value and anticipates possible material gain, one also perceives the psychological pain of potential loss.

The larger and more immediate the prospects for material gain, the greater the political power used to ensure and expedite exploitation, because not to exploit is perceived as losing an opportunity. Exploitation for immediate personal gain is often justified by the following notion: If I don't get it, someone else will, which implies that I might just as well have it as someone else because someone will get it. It is this sense of impending loss that one fights so hard to avoid. It seems more appropriate, therefore, to think of resources as managing humans than of humans as managing resources.[78]

According to the 1995 United Nations report *Global Biodiversity Assessment*, for example, the price of a fish dinner keeps going up in the United States as the North Atlantic is fished nearly dry. In the Amazon, cases of malaria are skyrocketing because clearcutting of the tropical rain forest is causing an explosion in the population of disease-carrying mosquitoes. In Southeast Asia, annual flooding is wiping out entire villages and even towns because the mangrove forests, which once buffered them from the floodwaters, have been reduced by half.[62]

Historically, then, any newly identified resource is inevitably overexploited, often to the point of collapse or extinction. Overexploitation is based, first, on the perceived rights or entitlement of the exploiter who discovered that which is to be exploited, to get his or her share before someone else does (the rights of ownership and private property). Second, overexploitation is based on the right or entitlement of the exploiter to protect his or her economic investment (again, the rights of ownership and private property).

There is more to it than this, however, because the concept of a healthy capitalistic system is one that is ever-growing, ever-expanding, but such a system is not biologically sustainable. With renewable natural resources, such nonsustainable exploitation is a "ratchet effect," where to ratchet means to constantly, albeit unevenly, increase the rate of exploitation of a resource.[11,78]

The ratchet effect works as follows: During periods of relative economic stability, the rate of harvest of a given renewable resource, say timber or salmon, tends to stabilize at a level that economic theory predicts can be sustained through some scale of time. Such levels, however, are almost always excessive, because economists take existing unknown and unpredictable ecological variables and convert them, in theory at least, into known and predictable economic constants in order to better calculate the expected return on a given investment from a sustained (but not biologically sustainable) harvest.

Then comes a sequence of good years in the market or in the availability of the resource, or both, and additional capital investments are encouraged in harvesting and processing because competitive economic growth is the root of capitalism. When conditions return to normal or even below normal, however, the industry, having overinvested, appeals to the government for help, usually with the coercive threat of eliminating jobs, because substantial economic capital is at stake. The government typically responds with direct or indirect subsidies, which not only encourages continual overharvesting but also privatizes the profits while passing the costs on to taxpayers.

The ratchet effect is thus caused by unrestrained economic investment to increase short-term profits in good times and strong opposition to losing those profits in bad times. This opposition to losing profits means there is great resistance to using a resource in a biologically sustainable manner because there is no predictability in yields and no guarantee of increases in yields in the foreseeable future. In addition, our linear economic models of ever-increasing yield are built on the assumption that we can in fact have an economically sustain*ed* yield. But this contrived concept fails in the face of the biological sustain*ability* of that yield.

Then, because there is no mechanism in our linear economic models of ever-increasing yield that allows for the uncertainties of ecological cycles and variability or for the inevitable decreases in yield during bad times, the long-term outcome is a heavily subsidized industry. This amounts to corporate welfare to maintain artificially high profits at taxpayer expense. It therefore pays industry to continually overharvest the resource on an artificially created sustained-yield basis that is not biologically sustainable. The flip side of outright subsidies is an economic "incentive" paid to a person or industry to induce them to "harvest" a renewable natural resource within biologically sustainable limits. Such incentives, however, are no more than *moral bribes* to do that which is biologically sustainable and therefore equitable in terms of future generations.

However, when the biological sustainability of a renewable resource is called into question in the face of industrial overexploitation, the industry marshals all scientific data favorable to the status quo as "good" science and discounts all unfavorable data as "bad" science. The stage is thus set on which science is politicized, largely obfuscating its service to society.

Because the availability of choices dictates the amount of control one feels one has with respect to one's sense of security, a potential loss of money is the breeding ground for environmental injustice. This is the kind of environmental injustice in which the present generation steals from all future generations by overexploiting a resource rather than facing the uncertainty of giving up potential income.

There are important lessons in all of this for anyone involved in sustainable community development. First, history suggests that a biologically sustainable use of any resource has never been achieved without first overexploiting it, despite historical warnings and contemporary data. If history is correct, resource problems are not environmental problems but rather human ones, which we have created many times, in many places, under a wide variety of social, political, and economic systems.

Second, the fundamental issues involving resources, the environment, and people are complex and process driven. The integrated knowledge of multiple disciplines is required to understand them. These underlying complexities of the physical and biological systems preclude a simplistic approach to both the use of renewable natural resources and conflicts resulting from human greed. In addition, the wide natural variability and the compounding cumulative influence of continual human activity mask the results of overexploitation until they are severe and often irreversible.

Third, as long as the uncertainty of continual change is considered a condition to be avoided, nothing will be resolved. However, once the uncertainty of change is accepted as an inevitable, open-ended, creative process of life, most decision making is simply common sense. For example, common sense dictates that one would favor actions that have the greatest potential for reversibility, as opposed to those with little or none. Such reversibility can be ascertained by monitoring results and modifying actions and policies accordingly.

Corporate interests are not vested in the reversibility of potential ecological disasters when they in fact are brought about by corporate greed and irresponsibility in the realm of the global commons, which is the biological sustainability of our biosphere. The current opiate of the masses in the arena of the global economy is the "free" market, which commonizes costs—payable by the consumer—and privatizes profits receivable by the economic elite. As French author and critic Anatole France observed: "If fifty million people say a foolish thing, it is still a foolish thing."

The Power of Economics and the Economics of Power

To the corporate mind, control is tyranny, whereas being out of control is freedom.[80] An economy that is out of control is a "free" market. But there is no such thing as absolute freedom; freedom is relative and always within restraints. The question thus becomes: what sort of restraints do corporations accept? Are corporations other-centered in their interests and therefore ethics-centered in their operations, or are corporations self-centered in their interests and therefore self-serving in their operations?

Corporate freedom relates to corporate power, and absolute power corrupts absolutely. Therefore, while corporations want absolute freedom from the "tyranny" of outside controls, which are in the public favor, they want to impose the tyranny of corporate control in politics in order to secure corporate control of the global market, which is in the corporate favor.[80]

As Professor Noam Chomsky puts it: "Power is increasingly concentrated in unaccountable institutions. The rich and powerful are no more willing to submit themselves to market discipline or popular pressure than they ever have been in the past."[79]

In considering democracy, Chomsky points out that the power is continually shifting away from parliamentary institutions into the hands of huge transnational corporations. Power is flowing to corporations and their supporting structures, all of which are completely unaccountable. The corporation itself has a stricter hierarchy than any human institution, which Chomsky says is a form of totalitarianism and unaccountability that is tantamount to "economic fascism." This is the reason corporations are so strongly opposed to classical liberals.

Thomas Jefferson, who lived just long enough to witness the early development of the corporate system, saw the handwriting on the wall. He said in 1799 that "Banking establishments [his term for corporations] are more dangerous than standing armies." Jefferson warned in his last years that money and banking establishments (corporations) would destroy liberty and restore absolutism, effectively nullifying the ideals for which the American revolution was fought.[79]

Preoccupied with consumption for its own sake, economics, since Jefferson's time, has increasingly been confined to the shallowness of appearances. Economists remain mesmerized by the theory that more rational analysis of the material world can and will provide all the necessary answers, despite the acknowledged limitations of "objective" reasoning in the physical sciences, says Frances Hutchinson.[81] "Economics as practised by professionals is, indeed, the 'dismal science' from which life itself is banished." She asserts that economists hide behind a "smoke-screen [they] erected...to obscure the fact that they know nothing about the real world."

The marketplace satisfies only temporarily our collective neuroses, while hiding the values that give meaning to human life. Although one's work can offer intense personal satisfaction, the economist has no mechanism to register the joy (utility) derived from work, which leaves the economist incapable of recognizing the fulfillment rendered by labor. Therefore, "environmental economics," which attempts to place monetary value on the nonmarketable aspects of both culture and Nature, dwells in the same murky habitat as marketplace economics.[81]

Gross materialism and gross poverty are equally ugly, and both are the result of an economic system that reduces everything of value to the cash nexus. When, for example, labor becomes a mere commodity, subjected to the laws of supply and demand, statistics substitutes for real human value. "People become things, children born and unborn are material possessions, a beautiful landscape has a price, and all are disposable in the name of economic rationality."[81] And thus, sadly, Jefferson's foresight comes to pass.

Today, the social benefits of knowledge—at times even stolen knowledge, and hence stolen benefits—are subjected to the greed of material hoarding for monetary gain. And greed is a nationality all its own. Consider that a central part of both the General Agreement on Tariffs and Trade and the North American Free Trade Agreement is protection for the ownership of knowledge and technology, which is rights of intellectual property.[82] The purpose of these agreements, says

Chomsky, is to ensure that the technology of the future is held in monopoly by huge private corporations, which are usually government subsidized.

Chomsky's point is a good one, because the wealthy industrialized nations, having greatly reduced the biological and genetic diversity within their own borders, are now looking to the nonindustrialized countries for genetic resources and traditional medicinal plants. For example, Canadian varieties of wheat contain disease-resistant genes from fourteen poorer nonindustrialized countries, and American cucumbers rely on genes from Korea, Burma, and India.[83]

Less wealthy, less industrialized countries, such as Brazil, India, and the Congo, still have relatively untapped gene pools for food plants and animals and medicinal plants of potential economic importance, but do not have the financial resources to commercialize them. Take the neem tree. Its bark, oil, and gum have been used for centuries by India's peasant farmers as a source of pesticides to protect their crops. Today, the neem tree, regarded as sacred by many Hindus, is the focus of an international dispute over who should control, and profit from, the tree's biological properties, an industrialized country or India, where the tree grows.[83]

I heard much about the neem tree and the controversy surrounding it at a conference on native medicinal plants while I was in Malaysia. The point is that the wealthy industrialized countries have the financial capital to study, identify, purify, and commercialize the biological properties of the neem tree, but they do not want to share any of the profits with the country where the tree grows or with the people who long ago learned how to use it. Such naked greed and arrogance are nothing less than intellectual pirating of traditional indigenous knowledge from people who have no voice with which to speak for themselves and cannot afford legal representation—and should not have to!

"If the Western scientists and multinationals really want to help developing countries, such as India, they should share their knowledge and shouldn't patent material derived from the genetic resources which those countries possess," says Ashish Kothari, a professor at the Indian Institute of Public Administration. This prompted Kamal Nath, India's former minister of the environment, to say: "What we protect becomes the raw material for the biotechnology industry of the developed countries without any benefits accruing to India."[83]

He went on to say that India is drafting legislation to regulate the removal of such material and may link that regulation to the issue of stronger safeguards for "intellectual property rights" demanded by the United States.[83]

Another example involves Jamaican peasants who use pink-flowered periwinkle to treat diabetes. When scientists grasped the meaning of an opportunity to use the properties of this plant for medicine, they extracted two drugs from it, the annual sales of which exceed $100 million. In the United States alone, $6 billion is spent annually on medicines derived from tropical plants. And the as yet untapped potential is spectacular; for every plant that has been examined by Western science, 200 more await study,[84] which spurs the naked greed of corporations.

Such corporate theft as the rights of intellectual property is increasing and is just one more example of corporate coercive power, in this case through the control of medicines and foods.

And speaking of food, the small, diversified family farm in the United States is being gobbled up by huge corporations, which in turn control the production, preparation, distribution, and therefore the pricing of food. Since 1930 the number of farms has declined steadily from six million full-time farms to less than one million in 1994. Compared to other businesses, farm debt has risen, and farming is now a more hazardous occupation than mining, according to John Kinsman, a dairy farmer in LaValle, Wisconsin.[85]

"Why protect family farms?" asks Kinsman. Because historically, and probably before, the family farm has been society's primary endeavor both socially and economically, and it has nurtured us physically, emotionally, spiritually, and economically. Communities and villages became possible through the enduring stability of the family farm. Democratic government grew directly out of such agrarian necessities and values as patience, cooperation, coordination, and shared responsibilities.[85]

Compared to corporate factory-style farming, the traditional family farm is far more environmentally friendly and benign. "Family farmers," says Kinsman, "have a vested interest in food, land, water and air quality because they actually live on the land they own, breathe the air, eat the food, drink the water and depend on continued soil productivity for future generations." It is a tragic mistake, present and future, to turn the responsibilities of trusteeship for family farms over to absentee owners who amalgamate them into huge corporate farms, which they see merely as a business, instead of a way of life in which they have a vested interest. I know this from firsthand experience.

Three companies in the United States process more than 80 percent of our beef, four companies control 50 percent of the hog production, and four companies oversee 95 percent of the chicken production. In addition, four corporations control 90 percent of the trade in grain on the world market.[85]

Although economists claim that the efficiencies of large-scale agribusiness are good for consumers, low farm prices do not translate directly into lower retail prices. For example, people who eat beef are probably unaware that the people who raise the beef have not received more than 75 cents per pound since December 1993 and have gotten less than 66 cents per pound since May 1995. But the meat packers have not passed this price on to consumers and have made unprecedented profits.[85]

The price paid to farmers for raw milk has fallen 15 percent since 1980 while retail prices have risen 15 percent. The extra profits are used by the corporate dairy industry to intensify its market presence through mergers and leveraged buyouts. Meanwhile, depressed prices for dairy products have in the last 15 years (1980–1995) forced more than half of the dairy farmers out of business. While family farmers have virtually no history of consumer blackmail by withholding production to exact higher prices, can the same be said for the corporate agribusiness?[85]

And in my own backyard, the corporate timber industry dominates the Coast Range mountains of western Oregon. Tax records for eleven "timber" counties in western Oregon show that 59 percent of the private timberland is controlled by ten corporate owners, according to a 1995 study by the Coast Range Association of Newport, Oregon. In Benton County, for example, 595 owners control 1,906 acres (1.5 percent) of timber as opposed to 120,137 acres (98.5 percent) controlled by ten corporate owners. In neighboring Lincoln County, 865 owners control 1,763 acres (almost 2 percent) as opposed to 105,931 acres (98 percent) controlled by ten corporate owners.[86]

Thus, in those two counties alone, 1,460 small owners control 3,669 acres (less than 1 percent) compared to 400,152 acres (more than 99 percent) controlled by 15 corporate owners. Out of these 15, 4 national and/or international corporations own 221,002 acres (more than 50 percent) of the timber land.[86] In addition, corporate timber owners have a history of shipping unprocessed logs to Japan, where they get larger profits than they would in domestic markets, which bypasses domestic mills and domestic workers.

At the same time, more than half of the large streams, which are critical habitat for endangered stocks of steelhead trout and commercially important salmon, run through private lands. "The fate of coastal salmon lies in the hands of corporate landowners," says Chuck Willer, director of the Coast Range Association. "Many industrial owners…are not at the table and are avoiding hard questions about their land's condition."[86]

"Communists and capitalists are alike in their contempt for country people, country life, and country places," observes author Wendell Berry. "They have exploited the countryside with equal greed and disregard. They are alike even in their plea that it is right to damage the present in order to 'make a better future.'"[59]

"This," says Chomsky, "is the model for the free-market future. The profits are privatized and that's what counts—it's socialism for the rich: the public pays the costs and the rich get the profits. That's what the free market is in practice,"[79] and it will surely continue to destroy the communities and cultures of the world until governments, including that of the United States, become accountable to their people. But as Frederick Douglass pointed out so long ago: "Power concedes nothing without a struggle." Mahatma Gandhi, on the other hand, noted that "even the most powerful cannot rule without the cooperation of the ruled."

It seems, after all, that Edmund Burke's admonishment was correct:

> Men are qualified for civil liberty in exact proportion to their disposition to put moral chains upon their own appetites.…Society cannot exist unless a controlling power upon will and appetite be placed somewhere, and the less of it there is within, the more there must be without. It is ordained in the eternal constitution of things that men of intemperate minds cannot be free. Their passions forge their fetters.[37]

The Economy of Community Is the Economy of Human Dignity and Social/Environmental Sustainability

James Madison once admitted that the Constitution of the United States was designed to thwart democratic rule. He argued in *The Federalist Papers* that the structure of the government of the United States, which replaced the Articles of Confederation, would "make it less probable that a majority of the whole will have a common motive to invade the rights of the minority" or, were such a motive to develop, make it "more difficult for all who feel it to discover their own strengths and to act in unison with each other."[87]

Madison was concerned with the wealthy minority, of which he was a part, and not the 116 African slaves he owned. He was worried about the property owners and businessmen who might lose some of their wealth and power at the hands of a democratic legislature. These biases still shape governance in the United States, and they still provide corporations with the institutional edge to sabotage proposed reforms or to sidestep those enacted. Reforming the legal/institutional framework of governance and commerce would require great personal courage and substantial political will to make the necessary structural changes in law and policy that would protect the reforms from the coercive demands of corporate power. The goal would not be to replace capitalism but to save it from itself.[87]

What must be done to save capitalism from itself? If capitalism is to survive, it must in fact become democratic in the best sense of the term. Václav Havel, president of the Czech Republic, has sound counsel for contemporary humanity. If democracy is to survive, expand successfully, and resolve the ever-present conflicts amongst cultures, says Havel, it must rediscover and renew its own transcendental origins.[88]

The very things for which Western democracy is most criticized, such as loss of morality, the crisis of authority, and the reduction of life to the self-centered pursuit of immediate material gain, do not originate in democracy but in the loss of spirituality, which is the only genuine source of responsibility and self-respect. Because of this loss, democracy is losing much of its credibility.[88]

All the tenets of democracy are merely technical instruments, which enable people to live in dignity, relative freedom, and responsibility; they cannot in and of themselves guarantee human dignity, freedom, or responsibility. The source of these basic human potentials lies in our relationship to that which transcends us—our spirituality.[88]

The effective expansion of democracy, not only from neighbor to neighbor but also internal to our individual selves, presupposes a critical self-examination. Internalizing democracy is the best way to save today's global civilization as a whole, including capitalism, not only from the danger of a conflict of cultures but also from the many other dangers that threaten it from within.[88]

The internalization of true democracy was central to Gandhi's vision of self-sufficient but interconnected village (community) republics with sustainable small-scale economic structures based on participatory democracy. E.F. Schumacher, who described Gandhi as the "People's Economist," was one of the rare Western economists to both understand and appreciate Gandhian economics. The following discussion is based on an article by Surur Hoda.[89]

To Gandhi, political freedom was merely the first step toward the real independence of village communities. "If the villages perish, India will perish," he said. The same is true for the United States or any other nation. Gandhi had his finger on the pulse of the masses and "refused," in Schumacher's words, "to treat economics as if people did not matter."

Therefore, when someone told him that no religion was good that did not make sense in terms of economics, he countered that no economics was good that did not make sense in terms of morality. The crossroad we face today in economics thus becomes clear: economic reasoning based on people (Gandhian economics) versus economic reasoning based on goods (Western corporate economics).

As Schumacher studied Gandhian economics, he found the most disturbing aspect of "developing" countries to be the millions of people without work. The question of how these millions could be helped to help themselves prompted Gandhi to call for "production *by* the masses" instead of "mass production." "The salvation of India," he said, "is impossible without the salvation of the villages [communities] and their inhabitants."

Production by the masses equates with sustainable community development just as surely as mass production equates with the current global demise of communities, especially rural communities, and the environment as a whole. Economic reasoning based on mass production of goods ignores, with equal ease, both people and the environment on which they depend for life.

Economic reasoning based on mass production of goods is concerned solely with increasing the supply of goods by means of advanced technology. According to such reasoning, industries should be large scale, capital intensive, and labor saving, even to the point of total automation to eliminate people as much as possible.

From the "goods" point of view, people are seen as liabilities to both production and profits because they make mistakes, are not punctual, argue back, join trade unions, and demand paid vacations, sick leave, medical benefits, and retirement pensions. Machinery does none of these things. The ideal, therefore, is to eliminate the human factor, something corporate industry is well on the way to doing—and in the process is destroying communities worldwide.

Sustainable community development is dependent on the five pillars of Gandhian economic thinking, which Schumacher identifies as follows: nonviolent, simple, small, capital-saving, and rural-based.

Nonviolent—Gandhi perceived modern industrial civilization to be exploitative in terms of the environment, such as massive clearcut logging, nonsustainable grazing by domestic livestock, strip mining, and commercial fishing to the point of species extinction (of which the seahorse is a new victim[90]). Destructive exploitation of the environment is violence against both the environment and people, especially the generations of the future. But people are also exploited directly, as in child labor, nonsustaining wages, and automation to eliminate jobs. "It would strip the world like a locust," Gandhi said of modern industrial civilization—and it is. At risk are biological, cultural, and spiritual wealth.

Simple—Gandhi's hallmark was a consciously simple lifestyle based on a rich spirituality. His sole possessions at his death were a pair of slippers, a watch, a pair of glasses, and a few loincloths. He lived in this conscious simplicity by choice for most of his life, in keeping with his motto: "Simple living, and high thinking."

Gandhi's notion that "high thinking is inconsistent with complicated material life" does not mean that conscious simplicity must equate with material austerity. Conscious simplicity is the lifestyle of sufficiency and joy predicated on an inner phenomenon. Material austerity, on the other hand, is a standard of living based on a sense of lack of material possessions predicated on an outer phenomenon. Thus, there are two ways to wealth: to want less or to work more. When compared to conscious simplicity, therefore, austerity is foreign territory.

Small—When Gandhi talked about "not mass production but production by the masses" or when he said that "production and consumption must be reunited," he was pointing out that the greater the size of the production unit, the greater the separation in distance and human relationships between production and consumption. Reuniting production and consumption is only possible if production units are small enough to be easily adapted to local needs in local communities. Another enormous advantage to both small-scale production and consumption is the minimization of transportation, which in itself adds substantially to the cost of goods but adds nothing to their real value.

"Bring work to the people and not people to the work," said Gandhi, which caused Schumacher to ask: "Can we utilize science and technology to this end?" Of course we can; it is only a choice. Mini-production plants can be developed so that people who live in small communities can once again become productive, but without having to depend on people already rich and powerful to provide jobs that only disappear due to automation and downsizing.

Capital-Saving—Although capital-saving is one pillar of Gandhian economics, the world is rushing with ever-increasing speed into large-scale, capital-intensive production. This is leading society into a crisis of survival, first by making people "machine minders," which does nothing to develop their humanity and robs them of their creative powers, and then by eliminating their jobs. Highly capitalized technology has proven monstrously inefficient and ineffective in solving the human/environmental problems of the world.

In addition, highly capitalized technology stimulates the ratchet effect in resource exploitation. As mentioned earlier, ratchet effect economics generally includes corporate welfare through subsidies, which translates into corporate profits through an immediate monetary expense to taxpayers and a long-range ecological/monetary expense to future generations.

Highly capitalized technology can also be subtle in its destructive power. Consider, for example, extinction of domestic animals, such as the Taihu pig, gembrong goat, and choi chicken.[91] These domestic Asian animals, along with as many as 1,500 other farm breeds worldwide, are as endangered as their wild relatives. The demise of biological diversity on the farm could prove equally damaging to the loss of biological diversity in the wild.

Causes for the decline in farm breeds, according to the U.N. Food and Agriculture Organization, include loss of habitat due to growth in the human population and wars. But the greatest threat comes when farmers in poor nations discard native breeds and switch to Western commercial livestock, which is highly productive.

Western breeds like Holstein cows, Rhode Island red chickens, and Yorkshire pigs are alluring to Asian farmers because of the great quantity of milk, eggs, and meat they produce. For a poor farmer barely able to feed his family, they seem heaven-sent. The problem is that Asian farmers invariably cannot afford the high cost of maintaining and feeding such specialized Western breeds.

Livestock that are bioengineered in Western laboratories for the technology- and money-intensive agriculture of Western industrialized countries may not be suited to other environments, cultures, or methods of farming. Local breeds may well prove better and more profitable in the long run than those marketed by the West.

Local breeds are the result of successful adaptations to particular environments that began when people started domesticating animals more than 10,000 years ago for food, fiber, the power to work, and for their droppings as fertilizer. The following are examples of such adaptations: China's min pig tolerates extreme temperatures, the pygmy hog of northern India is ideal for small villages, and the zebu cattle of Java are disease resistant and prolific. But now 105 such domestic animals are endangered in Asia, and Asia is not alone.

The U.N. Food and Agriculture Organization estimates that 30 percent of the 4,000 to 5,000 breeds of domestic animals thought to exist in the world are threatened with extinction and that three breeds become extinct every two weeks. Half of all the domestic breeds that existed in Europe at the beginning of the 1900s, for instance, have vanished, and more than one third of all breeds of poultry and livestock in North America are rare, which may well have a tremendous impact on rural economies in the future.

Rural-Based—To Gandhi, political freedom was only "the first step" toward social, moral, and economic independence of the community life of villages. In a document which has become known as Gandhi's "Last Will," he told his followers in the Sarvodaya (Welfare of All) Movement for the uplifting of India's villages: "You cannot build non-violence on a factory civilization. The economy which I have conceived eschews exploitation, because exploitation is the essence of violence. You have to be rural-minded before you can be non-violent...."

City-based industries both destroy the productive capacity of rural people and rob them of their livelihood. The loss of their livelihood causes rural people to migrate to the cities in search of work, which usually adds to a city's slums and to its roll of unemployment welfare. Only by improving the quality of life and creating work in the rural areas can the tide of human migration from rural communities into already overcrowded cities be stemmed or reversed. Rural communities must therefore be revitalized and revalued socially.

This revitalization and revaluation was behind Gandhi's Sarvodaya Movement. In the Sarvodaya Movement, said Schumacher, the village would become the basic unit of the economy, in which agriculture would remain the basic industry. In addition, however, other small-scale industries would be developed through technology based on social/environmental sustainability. "If we tap all our resources," said Gandhi, "we can again be rich, which we were at one time. We can repeat the phenomenon if we profitably occupy the idle hours of the millions."

Schumacher offered the following example of rural economy in India to illustrate Gandhi's point about occupying the "idle hours of the millions":

> One of the greatest teachers of India, Lord Buddha, included in his teaching the obligation of every good Buddhist to plant and see to the establishment of one tree every year for five years running. This in five years would give 2,000 million trees. The economic value of such an enterprise, intelligently conducted, would be greater than anything promised by five-year plans. It would produce foodstuffs, fibres, building materials, shade, water—in fact almost

anything that is really needed. And all this could have been done without a penny of foreign exchange and very little investment.

Gandhi fought all his life to improve the lot of India's poor, and in so doing saw economic reasoning based on people—rather than on consumer goods—as the way to social/environmental sustainability in the villages of rural India and the world. To this end, he gave what is known as "Gandhi's Talisman" to his followers:

> Whenever you are in doubt, apply the following test: recall the face of the poorest and the weakest person you may have seen, and ask yourself if the step you contemplate is going to be of any use to that person. Will that person gain anything by it? Will it restore that person's control over his or her own life and destiny? In other words, will it lead to the well-being of the hungry and starving millions? Then you will find your doubts melting away.

Today's corporations seem the antithesis of Gandhi's Talisman. Their sole motivation appears to be "profit over all" or "profit at any cost." For example, I attended a conference on the environment in Slovakia in June 1992. Also in attendance was an American owner and president of a company that made plastic sheeting used to line garbage dumps in order to prevent toxins from leaking into the soil. He spent his allotted speaking time trying to sell his product to the Slovakians.

At one point, someone in the audience asked him if his company had a program to teach people about recycling.

"Why would I do that?" he asked. "I make my money from waste. I need more and more waste to make money."

At this point, I turned to my Slovakian host and said, "This man makes me sick. He's the ugly American; he's what the American dream has led to."

"Don't listen to him," said my host. "I'm not."

"I can't help it," I replied." "He's an American speaking English, and unfortunately he speaks like corporate America."

But suppose Gandhi's Talisman was hung in every corporate boardroom, and its essence became the motivation for all corporate decisions. Would sustainable community development become a possibility throughout a more peaceful world?

If Gandhi's Talisman became the motivation behind corporate thinking, then his prescription for a human-based economy for India, as summed up by Schumacher, would be applied to the world as a whole:

> 1. Start all economic reasoning from the genuine needs of the people and help the poor to help themselves out of poverty.
>
> 2. Revitalize and foster not only agriculture but also non-agricultural activities in the rural areas, such as cottage industries for potters, weavers, shoemakers, carpenters, blacksmiths, etc.
>
> 3. Resist the further concentration of the growing population in large cities by reversing the trend of migration from rural to urban areas.
>
> 4. Develop systematic policies, based on the best available knowledge, for the mobilization of all productive resources, the greatest of which is the people themselves.

Wendell Berry, thinking along similar lines, said that "we must ask how a sustainable local community might function." He then listed the following 17 rules,[59] some of which I have modified slightly:

> 1. Always ask of any proposed change or innovation: What will this do to our community, present and future? How will this affect our common wealth, present and future?
>
> 2. Always include local nature (land, water, air, plants, and animals) within the membership of the community.
>
> 3. Always ask how local needs might be supplied from local sources, including the mutual help of neighbors.
>
> 4. Always supply local needs *first.* (And only then think of exporting excess products, first to nearby cities, and then to others.)
>
> 5. The community must understand the ultimate social/environmental folly of the industrial doctrine of "labor saving" if that implies work of poor quality, unemployment,

exploitation (either human or environmental), or any kind of pollution or contamination.

6. The community must develop appropriately scaled value-added industries for local products if it is to be anything more than a colony of the national or global economy.

7. It must also develop small-scale industries and businesses to support the local farm and/or forest economy.

8. It must strive to produce as much of its own energy as possible.

9. It must strive to increase earnings (in whatever form) within the community, and decrease expenditures outside the community.

10. Money paid into the local economy must circulate within the community for as long as possible before it is paid out.

11. If it is to last, a community must be able to afford to invest in itself: it must maintain its properties, keep itself clean (without dirtying some other place), care for its old people, and teach its children.

12. The old and young must take care of one another. The young must learn from the old, not necessarily and not always in school. There must be no institutionalized "child care" and "home for the aged." A community knows and remembers itself by the association of its old and young.

13. Costs now conventionally hidden or "externalized" must be accounted for. Whenever possible, they must be debited against monetary income.

14. Members of the community must look into the possible uses of local currency, community-funded loan programs, systems of barter, and the like.

15. They must always be aware of the economic value of neighborliness—as help, insurance, and so on. They must realize that in our time the costs of living are greatly increased by the loss of neighborhood, leaving people to face their calamities alone.

16. A rural community must always be acquainted with, and thoroughly interconnected and interactive with, community minded people in nearby towns and cities, and vice versa.

17. A sustainable rural community will be dependent on the urban consumers' loyalty to local products. We are thus talking about an economy that will always be more cooperative than competitive.

A community economy, says Berry, is one whose aim is generosity and a well-distributed, well-safeguarded abundance rather then an economy in which well-placed persons can make a financial "killing." He suggests starting a "local food economy," by which he means local producers growing as much food as possible for the local economy, and the local people buying as much of their food as possible from local producers. Such food is better, safer, and fresher because it is grown locally.

The more local the food is, the fresher it is. And as the distance between producer and consumer is shortened, the consumer's power to know about the quality of the food and to influence that quality is increased.[92]

Although locally grown food may be less expensive because the costs of transportation, packaging, refrigeration, and national advertising campaigns have been eliminated, this is not always so. The price of a given food, say carrots, may be influenced by the scale on which it is produced somewhere else. Carrots mass produced in another state, for example, may actually sell for a lower price when compared to locally grown carrots, but the out-of-state carrots will likely lack the quality of those which are locally grown. Local carrots are more likely to be selected for flavor and nutritional value, whereas out-of-state carrots are more likely to be selected for their ability to be stored over long periods so they can withstand shipping.

Starting a local food economy is attractive because it is decentralized, can be small and inexpensive, requires no one's permission, and can involve everyone. "Of course," says Berry, "no food economy can be, or ought to be, only local. But the orientation of agriculture to local needs, local possibilities, and local limits is simply indispensable to the health of both land and people, and undoubtedly to the health of democratic liberties as well."

Berry goes on to say that society as a whole has much to gain from local land-based economies, which foster the idea of a culture's local adaptation within itself and with the ecosystem in which it rests. Such an economy would protect the sustainability of its resources, which, after all, are the energy that fuels the economy.

Such a food economy could be started in Asia, for example, where rice is in short supply. Although the production of Asian rice doubled through the irrigation of about 187 million acres of land between 1966 and 1985, the International Rice Research Institute, based in Manila, Philippines, says that yields have lagged behind the growth of the population for almost 20 years.[93]

Adding to the problem of shortages in rice is a looming shortage of water. In Asian countries, where water has always been abundant, severe problems will occur by 2025 because per capita availability of water has already decreased by 40 to 60 percent between 1955 and 1990. A total of 86 percent of all water used in Asia goes to agriculture, mainly the growing of rice. Because of impending shortages of water, new quickly growing varieties of rice will be developed to allow farmers to avoid late droughts or, where there is a long monsoon season, to grow two crops without irrigation.[93]

To stave off severe shortages of food by the year 2020, more than 350 million tons of rice must be produced, because rice is the staple food for half of the world's population, and more than 90 percent of it is grown and consumed in Asia.[93] Growing more than 350 million tons of rice could be used to revitalize the original food economies in Asia.

Consider that the rice once grown in local paddies was also hand pounded and winnowed by local people for local consumption, and the surplus was marketed in the nearest area where there was a shortage. Now, much, if not most, of the rice is grown with the aid of machinery, and the harvest is taken to rice mills in the large cities, where it is processed before being sent back to the villages.

Meanwhile, the villagers have lost their jobs. In addition, the rice at city mills comes from all over the country, and some of it is infected with location-specific diseases. But now the disease, which was originally "exported" from the paddy where the rice was grown to the mill, is exported from the mill to other paddies in other villages.

The opportunity at hand is to help the villagers learn about improved strains of rice and perhaps improved ways to grow them. Then, instead of shipping the harvest to city mills, paddy-thrashing equipment could be introduced into the villages and the villagers taught how to use it. If necessary or feasible, this could even be done on a cooperative basis with two or more villages sharing the equipment.

In this way, the villagers would again have work, and the dignity that comes with it. They could feed themselves and one another directly at less cost, because the cost of transportation to and from the city mills would be eliminated, as would that of the rice brokers. The people could grow what they actually need and want and in the process look out for one another and their neighbors.

In addition, there would be no outside corporation telling the villagers what they can and cannot do with their own rice paddies and rice harvests, and any potential corporate influence in setting or maintaining artificially high prices for rice would be eliminated. Finally, the village as a whole would have an incentive to care for its lands because the sustainability of the village's economy depends on them.

Local Community within the Context of Landscape

A community's world view defines its collective values, which determine how it treats its surrounding landscape. In turn, the surrounding landscape is the community's social mirror. As the landscape is altered through wise use or through abuse, so are the community's options altered in like measure. A community and its landscape are thus engaged in a mutual, self-reinforcing feedback loop as the means by which their processes reinforce themselves and one other.

It is therefore imperative that local communities not only consider but also involve their surrounding landscapes in all aspects of sustainable development. Unfortunately, absentee owners (especially internationals) are seldom personally vested in the welfare and sustainability of the community or the landscape in which they have economic investments. Their allegiance, as previously stated, is to the profit margin, not to the community as a sense of place. So they are willing, more often than not, to break all the rules of sustainability in the name of maximizing short-term monetary gain.

It is this allegiance to the profit margin that is destroying plant and animal species at an "alarming rate." Over 30,000 species are threatened with extinction, according to the *Global Biodiversity Assessment* of the world's fading biological and genetic diversity. As a consequence, deadly diseases are going untreated all over the world because natural medicines, which could have saved lives, have simply vanished before their benefits could be discovered. In addition to the increasing number of endangered species, the report chronicles the loss of genes, habitats, and even entire ecosystems.[62,94]

"Biodiversity," according to a summary of the report, "represents the very foundation of human existence. Yet by our heedless actions, we are eroding this biological capital at an alarming rate."[94]

The report cites an example of protecting such seemingly unimportant species as the day-flying moth, *Urania fulgens*, found in Mexico and South America. The moth metamorphoses from a caterpillar that feeds exclusively on a particular variety of trees and vines known as Omphalea.

The heavy defoliation caused by the feeding of the caterpillars in turn causes the plants to produce a protective chemical toxin, which makes them unpalatable to the moths. The plants' toxic compounds have been effective against the AIDS virus in test-tube experiments. There is a caveat, however; the toxin is produced only when a plant interacts with a large population of caterpillars.

The report goes on to say that the loss of genetically distinct populations of known species poses consequences almost as serious as the loss of whole species. It then cites several reasons for the decline in both genetically distinct populations and species themselves: (1) increased population and economic development deplete biological resources; (2) human failure to consider the long-term consequences of actions that destroy habitat, exploit natural resources, and introduce nonindigenous species; (3) failure of economic markets to recognize and accept the true value of protecting and maintaining the variety of species; (4) increased human migration, travel, and international trade; and (5) the spread of pollution affecting the quality of air and water.

The lesson for local communities within this global scenario is that long-term sustainability is dependent on local biological and genetic diversity. Let's consider a couple of examples, both of which center around a community that depends on its surrounding forest for most of its jobs—a specialized community and a diversified community.

Specialized Community

The first example is that of a specialized community in the Coast Range of western Oregon whose economy is based almost solely on the extraction and milling of timber, primarily old-growth timber. The method of logging the timber is clearcutting, which removes everything. Once cut, most of the remaining large woody material in an area is burned to make planting the next crop of trees easier. When the areas are planted, the original diversity of trees is replaced with a monoculture of evenly spaced Douglas fir in neat rows, the purpose of which is to maximize their growth and minimize the time until the next clearcutting can take place.

The original forest is thus converted into an economic tree farm for the quickest possible economic gains at the least possible economic cost. The long-term price is the loss of the forest and its biological and functional diversity. What might this mean to the community?

First, the old-growth is cut as quickly as possible because any merchantable tree left to fall and rot, reinvesting its biological capital into the soil, is considered by the timber industry to be an economic waste. The old-growth trees are therefore cut as fast as possible, which means much faster than the replacement tree-farm crop can grow to the predetermined age for harvest. When, therefore, the community runs out of old-growth trees to cut, the mill shuts down, and all of the primary forest-related jobs are lost.

How fast a "timber" community runs out of timber-related jobs depends on whether the private forest land and mill are locally owned or in absentee ownership. If a local person owns the timber and the mill, then he or she would have a vested interest in keeping the mill operating as long as possible and may not cut the trees so fast. An absentee owner, on the other hand, may not only cut the trees as fast as possible but also ship them whole to some other country, which steals jobs from domestic mill workers. A local owner of timber (without a mill) might sell her or his timber to a mill owner somewhere else and therefore might also cut the trees much faster than if she or he owned a local mill, which needed a constant supply of timber.

In any event, the local forest is converted to an economic tree farm with some short-term and long-term consequences for the community. The first consequence, an immediate one, is a loss of local biological diversity, such as the plants and animals associated with the forest that cannot or will not exist in a tree farm, especially one with a short economically driven life cycle. In addition, the timber industry is interested only in the conversion potential of the trees (as opposed to their intrinsic value) and therefore uses herbicides and pesticides to eliminate as much biodiversity as possible in favor of making the crop trees grow as purely as possible and as quickly as possible. It also eliminates as much genetic diversity from the crop trees as possible by selecting those trees that grow the fastest.

In so doing, the industry converts a diversified forest into a specialized crop, which constrains how the community can use the land in the short term. Simultaneously, however, the timber industry also commits two or three generations of people living in the community to be largely surrounded by an economically designed landscape, which ultimately affects such things as species of plants and animals, soil, and water.

When, for instance, the old-growth forest has been liquidated, no more old trees will stand as living monarchs, to die and stand as large dead trees (snags) and to topple as large fallen trees and lie for centuries decomposing, providing a kaleidoscope of habitats, and performing their myriad functions as they recycle and reinvest their biological capital into the soil from which they and their compatriots grew. The forest, the standing large dead tree, and the large fallen tree, which are only altered states of the live old-growth tree, will go the way of the oldest living thing on Earth, the old-growth monarch of the forest, and become locally extinct.

With the extinction of the old-growth forest will go such species as the northern spotted owl and the marbled murrelet, which have evolved in concert with that particular habitat. In fact, the owl and the murrelet have adapted to particular features of that habitat.

The northern spotted owl nests in tall, broken-topped, old-growth Douglas fir trees. The marbled murrelet, a seabird, nests on carefully selected large, moss-covered branches at least 100 feet up in old-growth trees, with other branches close overhead to protect the nest site. The murrelet's nest tree is located several miles inland from the coast, where it feeds. Because they are so specialized in the selection of their reproductive habitats, neither the owl nor the murrelet is capable of adapting to the rapid changes wrought by the industrial liquidation of the old-growth forest.

An interesting twist to the story is that not only will species of plants and animals become extinct with the liquidation of the local old-growth forests; so too will the "grandparent trees." As young trees replace liquidated old trees in crop after crop, the ecological functions performed by the old trees, such as creation of the "pit-and-mound" topography on the floor of the forest, become extinct because there are no more grandparent trees to blow over.

The "pit" in pit-and-mound topography refers to the hole left when a tree's roots are pulled from the soil, and "mound" refers to the soil-laden mass of roots, called a rootwad, suddenly projected into the air above the floor of the forest. The young trees that replace the grand-parent trees are much smaller and are different in structure. They cannot perform the same functions in the same ways.

Of all the factors that affect the soil of the forest, the roughness of the surface caused by falling grandparent trees, particularly the pit-and-mound topography, is the most striking. It creates and maintains the richness of species of plants in the herbaceous understory and affects the success of tree regeneration.

Uprooted trees enrich the forest's topography by creating new habitats for vegetation. Falling trees create opportunities for new plants to become established in the bare mineral soil of the root pit and the mound. In time, a fallen tree itself provides a habitat that can be readily colonized by tree seedlings and other plants. Falling trees also open the canopy, which allows more light to reach the floor of the forest. In addition, pit-and-mound topography is a major factor in mixing the mineral soil and organic topsoil of the forest floor as the forest evolves.

Although young trees also fall, they have neither the height and volume of crown, the size of stem, nor the depth and mass of roots that grandparent trees have; they cannot, therefore, serve the same function. The extinction of the grandparent trees changes the entire complexion of the forest through time, just as the function of a chair is changed when the seat is removed. The "roughness" of the floor of the forest, which over the centuries resulted from the cumulative addition of pits and mounds and fallen grandparent trees, will become unprecedentedly "smooth."

Water moves differently over and through the soil of a smooth forest floor, one devoid of large fallen trees to act as reservoirs, storing water throughout the heat of the summer and holding soil in place on steep slopes. The huge snags and fallen trees that acted as habitats are gone, as are the stumps of the grandparent trees with their belowground "plumbing systems," which guided rain and melting snow deep into the soil.

This plumbing system of decomposing tree stumps and roots comes from the frequent formation of hollow, interconnected, surface-to-bedrock channels that drain water rapidly from heavy rains and melting snow. As roots rot completely away, the collapse and plugging of these channels force more water to drain through the shallower soil matrix, which reduces soil cohesion and increases hydraulic pressure, which in turn causes mass soil movement. These plumbing systems cannot be replaced by the young crop trees of tree farms.

Thus, by simplifying the composition of plant and animal species, the industry simplifies the structure of the forest, converting it into a tree farm. Such economically driven simplification extirpates many of the interactive, interconnected, interdependent functions on which the long-term stability and adaptability of a forest and its integrated human community depend.

If, however, a community whose specialized economy has collapsed turns lock, stock, and barrel to another single endeavor, such as tourism, it is still putting all of its eggs into one economic basket. It is just a different basket. To survive in a sustainable manner, a community must diversify within the context of its landscape.

Diversified Community

In the second example, a community, instead of becoming specialized in the singular sense, diversifies and uses the whole forest to meet its local requirements for employment. In this example, let's assume that a local person owns some of the forest land surrounding a community and also owns a small sawmill. By cutting his timber at a biologically sustainable rate, he can keep the mill running, at least most of the year. If there are other local owners of small acreages of forest land, they could help to supply the mill with timber. Beyond that, there are a host of other potential products from the forest, nontimber products.[95]

The nontimber products from the forest were used by the First Americans and the early European invaders to help sustain themselves in everyday life. As such, plants have played a more important role in the lives of the First Americans than did animals. Not only were they used for medicines, but when the men caught such food as salmon, the women made it into a complete meal by stuffing it with herbs, roots, berries, and so on.[95]

Unfortunately, much of the knowledge of how to identify and use the plants has been lost to most contemporary First Americans. One of the main reasons for this loss was the European corruption, and often purposeful destruction, of the people's ancestral culture. Nevertheless, a few older First Americans still remember much of their ancestral knowledge and feel that before we can save the forest, we must learn about its other uses, aside from lumber and paper.[95] Herein lies an opportunity.

If a community has amongst its members one or more local owners of forest land who actually treat the land as forest, then all of the necessary pieces—diversity of species, structures, and functions—are intact, which allows the forest to act like a forest as opposed to a commercially simplified tree farm. Then, even though some logging may take place, which can be done sustainably, it is possible to develop nontimber products from the forest. Development of nontimber products could help loggers and even mill workers through slow times by augmenting their incomes and/or helping owners of forest land to augment their incomes.[95]

The first formal company to market nontimber specialty forest products in the Pacific Northwest, for example, was the Kirk Company, established in Puyallup, Washington, in 1939. It marketed the foliage of evergreen huckleberry, salal, and sword fern to local retail florists and eventually to wholesale florists in California and Illinois. This kind of business expanded to boughs of cedar, tips of white pine, limbs of Douglas fir, Oregon grape, holly, red-tipped huckleberry, and juniper, which were harvested and marketed from September to May, when the plants were tender during the growing season.[95]

Eventually, to meet the demand in an economically feasible way, many of the wholesale houses set up a network of small buying houses throughout western Washington and the northern Coast Range of Oregon. This enabled the wholesalers to concentrate small loads from individual pickers into large truck-sized loads, which could then be transported to processing plants, a system still in use today. Always on the lookout for fresh opportunities, the pickers of evergreen plants continually tried new things, such as picking up empty cones of fir and pine off the floor of the forest as additions to the wreath and craft markets for Christmas greenery.[95]

The future of the nontimber forest industry will continue to grow and thrive as long as it remains flexible and is realistic in its expectations. In the last five years, new cooperatives have come into being, which care for forested land and make wreaths for large companies, as well as harvest, sort, and box floral greens. They also harvest and dry medicinal plants. It takes little effort to organize and start a cooperative, which gives real power and often joy to its members as they collectively change and direct their futures. Such work is labor intensive and therefore benefits from people working together.[95]

Now there are such nontimber ventures as picking, processing, and marketing mushrooms and medicinal plants. In these latter endeavors, however, a problem has arisen: overexploitation through interpersonal competition from outside of the community.

For any venture to be of value to a community, it must be community based. It must be kept biologically sustainable if it is to be economically sustainable. And it must be cooperative and coordinated so that individual greed does not destroy the sustainability of the biological resource in the present for the future.

In addition to harvesting a nontimber resource from the forest, the community could develop local ways and means of creating finished products to market within and outside of the community. It might even be possible to establish a network among neighboring communities for an integrated venture in which several products are manufactured and marketed from local raw materials.

It is even possible to set up a cooperative with neighboring First American communities in which First Americans could be paid as teachers by sharing their knowledge of plants and their uses and also by working members of the cooperative. As such, the First Americans would also have the expertise needed to help maintain the biological sustainability of the nonforest products. If the cooperative were justly established, it would be good for both communities—the one with the traditional knowledge and the other with the usable land base.

Another possibility is to develop classes in community colleges and/or vocational schools dealing with all the aspects of creating cottage industries in communities. Some of these classes could be taught by First Americans, thus beginning a permanent interface between cultures. There could also be local short courses, and even traveling short courses, where local teachers go to other communities to teach courses for the local area.

Or a bioregional teaching consortium could be set up to hold annual classes for a wide geographical area, dealing perhaps with more than one ecosystem. Such a consortium could cover all aspects of a cottage industry, from biological sustainability of the natural resources, to harvesting, processing, marketing, setting up cooperatives, keeping the financial books, and so on. County extension agents and 4-H clubs could also help develop the idea.

Although diversification, properly managed at the human level, places less demand on a given resource by sustainably using the entire ecosystem by season, members of a community can also stretch their limited ecological and economic resources even farther by practicing conscious simplicity. The idea of wanting less, working less, spending less, and living more simply challenges the typical American ideal of getting ahead, of having more and bigger material things.

Conscious simplicity challenges the notion of bigger is better and more is preferable to less. Wanting less demands less from the landscape, thereby keeping it more flexible and adaptable ecologically.

Conscious simplicity leads not only to a sense of self-control but also to a sense of being in control of one's personal destiny. It also frees up time to be more involved in quality relationships within the community, which builds the social capital that knits people together and reduces internal community conflicts. Conscious simplicity thus brings to mind two words: leisure and busy.

The Chinese character for leisure is composed of two elements, which by themselves mean open space and sunshine. Hence, an attitude of leisure creates an opening that lets the sun shine in. The Chinese character for busy is also composed of two elements, which by themselves conversely mean heart and killing. This character points out that for the beat of one's heart to be healthy, it must be leisurely.[97]

According to Brother David Steindl-Rast (a Benedictine Monk), we tend to think of leisure as the privilege of the well-to-do. "But leisure," says Brother Steindl-Rast, "is a virtue, not a luxury. Leisure is the virtue of those who take their time in order to give to each task as much time as it deserves....Giving and taking, play and work, meaning and purpose are perfectly balanced in leisure. We learn to live fully in the measure in which we learn to live leisurely."[96]

Conscious simplicity also equates with both the aforementioned Gandhian economics and the seventeen points of economics necessary for a sustainable community as outlined by Wendell Berry.

There is in fact a growing trend toward conscious simplicity. The Trends Research Institute, in Rhinebeck, New York, estimates that 4 percent of the 77 million baby boomers in the United States are already making a conscious effort to simplify their lives, and it estimates that by the year 2000, 15 percent of the baby boomers will have joined the trend.[97]

Another study conducted in 1994 by Yankelovich Partners, a research organization based in Norwalk, Connecticut, indicates that the attitudes of many Americans are changing. The American dream is being scaled down; attitudes are changing in the most basic aspects of life: homes, jobs, clothing, food, indeed the very fabric of well-being.[98]

Yankelovich's Marjorie Goldstein noted that many Americans no longer feel it is necessary to have everything, such as a new car, a new seasonal wardrobe, or an expensive vacation. Moreover, the Yankelovich Monitor, a study that has tracked the social values of American consumers since 1971, reveals that American concepts of success and accomplishment have changed drastically.[98]

In 1989, for example, 39 percent of the people surveyed said they considered owning an expensive automobile to be a measure of success, a figure which dropped to 23 percent in 1994. The numbers fell 10 percent in every category: having a successful investment strategy, staying in a luxury hotel, shopping at prestigious stores, wearing designer clothing, and having a "gold" credit card.[98]

The study simultaneously indicated that other values associated with a feeling of success and accomplishment have increased correspondingly. In 1991, for instance, 63 percent of those surveyed said that "being satisfied with life" was a measure of success and accomplishment. That figure jumped to 81 percent in 1994; "being in control of your life" rose from 57 percent to 78 percent; "being able to afford what's important to me" went from 57 percent to 74 percent. In one year alone, "having a good marriage" jumped from 62 percent to 74 percent when associated with a feeling of success and accomplishment.[98]

After five decades of nonstop consumerism, the values of many Americans are shifting from the tangible to the intangible, from real estate to relationships.[98]

By setting priorities, an estimated four million Americans have joined in what is described as the "voluntary simplicity movement." By some estimates, the number could triple by the year 2000, although not everyone in the movement will be there by choice.[98]

Many workers are in the simplicity movement, but not voluntarily, says Gerald Celente, director of the Trends Research Institute. Many jobs are being lost to automation and corporate downsizing, and most of these jobs will not return. "The world is changing," says Celente. "It's the end of the industrial age and the beginning of a global age. And America is no longer the country it used to be."[98]

As we approach the twenty-first century, Celente believes the voices of the disaffected will blend with those who have consciously chosen voluntary simplicity. The latter form "the spiritual voice" behind the movement, which emerged about a decade ago as an antidote to escalating materialism and overconsumption. Simplifying as an antidote to escalating materialism and overconsumption can bring into question the meaning and purpose of one's life, including community. Resolving this question is very freeing emotionally.[98]

"If your spouse is the most important person in your life and the marriage is the most important relationship, that's where your energy should be invested," says John Billinsky, Jr., a Mennonite minister and psychiatrist at Prairie View in Newton, Kansas. "'Establish priorities is my advice," he says. "People need to acknowledge, at least intellectually, that they will not live forever, and decide what is important to them. That's where they should be putting their energies, and they should devote their resources to those priorities."[98] (There are a number of publications available on simplifying your lifestyle.[99])

Whether or not one opts for conscious simplicity, one of the great tragedies of communal life is when economic competition over natural resources (direct or indirect, conscious or unconscious) destabilizes an otherwise sustainable system. If economic destabilization is internal to a community, it is strictly a community problem. But economic destabilization is often external to a community, in which case the problem may be resolvable within a broader geographical area.

Beyond Community

For a community to be socially and ecologically sustainable, it must simultaneously be economically sustainable, which means that communities must cooperate and coordinate within a broad, well-defined geographical area or bioregion. This seldom happens, however.

I spoke recently to a large group of mayors and other city officials from Seattle, Washington, all the way south to Eugene, Oregon. The group was working with, as it was explained to me, a corridor of social/environmental sustainability within the Puget Sound Trough of western Washington southward throughout the Willamette Valley of western Oregon.

They were including only the floors of the two large valleys, but not the forested slopes of the Coast Range of mountains, which forms the western flank of the valleys, or the Cascade Range, which forms the eastern flank. There is no chance of social/environmental sustainability, however, without including the water catchment for the entire area, because without a sustainable supply of water, sustainability is merely an academic exercise.

Bioregion is a critical concept if communities are going to practice the economics of social/environmental sustainability. Without a collective vision of sustainability within a well-defined bioregion, the communities are no more than economic colonies for the national and international economic markets.

The whole principle of colonialism is to exploit someone else's natural resources, shipping as much of the principle as possible, as fast as possible, to whichever market will pay the highest price for the plundered booty. Thus, the more communities rely on outside markets, either for import or export of goods and services and/or jobs, the more they become economic and political colonies, which progressively give up self-rule—and therefore democracy.

This means that a centralized national and international economy is good for the corporate elite in their bid to control global markets. In the process, however, the controlling corporations treat the rest of the world, from first-world nations to third-world nations, as a gigantic colony from which to enrich themselves at the cost of the huddled masses.

Each community's economic sustainability demands that only the ecological *interest* of a bioregion is marketed. But the centralized corporate economy is in a constant feeding frenzy as it gobbles up all the ecological *principle* of all the available natural resources it can get. The legacy of this continual enrichment of the already wealthy minority is an increasingly fragile, ever-more endangered environment.

Social/environmental sustainability is therefore dependent on a decentralized economic/political system of democracy if economic sustainability is to be achieved. Economic sustainability, in turn, is dependent on the cooperation and coordination of communities sharing a common vision of the greatest possible economic independence within the broad landscape of a well-defined bioregion.

Such economic independence—and with it the return of a free democracy—will not be easily wrested from corporate control. But it is possible if communities can find the moral courage and political will to stand united in the face of corporate tyranny, which manifests itself as the economic slavery of all generations.

EDUCATING FOR SUSTAINABILITY

Education is essential to the freedom of self-government. It must, therefore, be a local, state, and national priority of the highest degree. If it is not, then one is courting local, state, and national despotism.

Education, both as formal academic training and as the experiences one has in the journey of life, is absolutely necessary for sustainable community development. Formal academic training and the experiences of life come together to make one literate, which means to be a well-informed, educated person. Literate also means having the ability to read and write.

Literacy is thus the sum of one's ability to use language to share one's knowledge, intuition, experiences, values, and talents (=communication), without which democracy, and thus sustainable community development, is impossible. There are many facets to literacy, of which five are particularly important to sustainable community development: academic, environmental, economic, democratic, and community.

Of these, environment, economics, and democracy have already been discussed in some detail and will merely be put into perspective with respect to sustainable community development. Academic literacy will be discussed briefly, because much has already been said and written about formal schooling and need not be repeated here. The major focus will be community literacy, because community, especially sustainable community, is the great crucible into which all facets of literacy ultimately pour, mix, and bear fruit in some form. There is, however, no one answer; the truth of sustainability lies in the sum of the whole.

Academic Literacy

If we really believe in the inherent value of democracy and the freedoms it brings, we must invest in empowering education for all of the people, which assumes that we believe in the intrinsic value of each and every person. But some say we already spend enough money on education; others want to cut the budget.

In both cases, people have become preoccupied with relatively minor details because they fail to proceed from a basic frame of reference: A society is only as free as the quality of its education. Without such a basic frame of reference, it is impossible to focus on a fundamental issue without getting hopelessly lost amidst a plethora of confusing and isolating details.

A society held in ignorance is powerless to govern itself. If we do not have a well-educated public, military power can easily replace civil liberty. All you have to do is look around the world to see the threat is real.

These two alternatives are important to understand because the gap between the current quality of our educational system and the necessities of a free democracy is enormous. We need to begin a revolution on behalf of our national education so that once again the real power of the people can come to the fore: the power to combat the coercive greed of the corporate elite, the power to restore real democracy in community, and the power to reclaim those inner values that make human life worth living.

True, we need to master mathematics, science, and technical skills, teaching them widely and understanding them within the context of whole systems, but that is not enough. We must also teach the price, value, and processes of self-government. We must blend science with intuition, humility, and sociology; we must integrate politics and commerce with humanity and ethics; and we must infuse it all with the fluid connectivity of language and civility so that we may create a community in which it is safe to share, innovate, imagine, and have the courage to fail our way to success and sustainability.

For us humans to continue our spiritual/intellectual development, to achieve the success of social/environmental sustainability, we need to protect one aspect of our culture that we normally neglect: language. Perhaps one of the greatest feats of humanity is the evolution of language, especially written language, which made culture possible.

Language, which we seem to take for granted, is not something we generally think of as becoming extinct. And yet languages are disappearing all over the world, especially those of indigenous people that are spoken languages only.

Of all the gifts of life, language is one of the most incredible. I can, in silence, understand what I think you wish me to know when you write to me. And I can perceive what I think your thoughts are and ask for clarification when you speak to me. You speak and you write and you allow me to share a small part of you.

Through language, we can create, examine, and test concepts, those intangible wisps of human thought and imagination. Concepts can only be qualified, not quantified; only interpreted, not measured. And concepts can be reinterpreted hundreds or even thousands of years after they were first written.

Language is the storehouse of ideas. It allows each succeeding generation to benefit from the knowledge accrued by generations already passed. It is a tool, a catalyst, a gift from adults to children. By means of language, each generation begins further up the ladder of knowledge than the preceding one began.

One of the greatest values of language is that it allows us to search for truth and to strive for those ideals that we, as individuals, as communities, and as societies, perceive to be right and just. In this sense, language has become an imperative for the survival of human society, because the tenets of society are founded on language. We simply must understand one another if our respective societies are to survive.

In a broader sense, every human language—the master tool that represents its own culture—has its unique construct, which determines both its limitations and its possibilities of expressing myth, emotion, and logic. As long as we have the maximum diversity of languages, we can see ourselves—the collective human creature, the social animal—most clearly and from many points of view in a multitude of social mirrors. And who knows when an idiom of an obscure language, or a "primitive" cultural solution, or the serendipitous flash of recognition spurred by some ancient myth or modern metaphor may be the precise view necessary to resolve some crisis in our local community or in global society.

We humans live in two interpenetrating spheres: that of Nature and that of words. These places are central to how we understand ourselves and thus how we live. In a sense, we are created by them, and in the meaning we bestow on them, we in turn create them. Of these two spheres, that of Nature must take precedence, because the diversity of many of its facets is at stake.[100]

A widely spread false assumption about language is the notion of its neutrality, that it is an unambiguous set of symbols for representing the world, from which it is independent. This means that the words we use stand for things and events in the world and their syntax for how those things and events relate one to another.[101] However, language is only metaphorical of that which we cannot approach directly, the truth about the things and events in the world we can only intuit but neither touch nor pin down with certainty.

I say this because I once wrote a book about the natural history of an ancient forest[75] in which I tried to define two words: *ancient forest*. On the one hand, I failed miserably, and on the other I succeeded brilliantly. I failed because I could not define "ancient forest." I succeeded because I learned that I cannot define anything; I can only characterize it but clumsily and in a roundabout way.

Inherent in a would-be definition of ancient forest is the notion that universal Creation is open-ended, full of wonder, full of unknowns, and full of seeming digression in our human view of history. We in Western society are in love with "clean," straight lines, with quantifiable results, with final products. Creation is none of these. Creation is a process, the very existence of which is Creation. I could not, therefore, define "ancient," or "forest," or "tree," or "soil," or "life," or "death," and so on because each is a part of the other and each is in the process of becoming something else.

Language thus lends itself to contradictory interpretations and uncertain meanings while simultaneously relating to how we experience the world and give meaning to our experiences. This, in turn, means that language and our world view affect each other in an ongoing process of perception, construction, articulation, reproduction, and the legitimation of ideology. Although language has opened us humans up to the world and one another, thereby enhancing our experience of both, it is also the tool with which we have gradually closed ourselves off from Nature and one another, thereby altering our experience of both.[101]

The metaphorical nature of language is pervasive and is not just a linguistical phenomenon but one that also affects how we perceive, think, and act in the world. Because language, thought, and action are intimately linked, a metaphor acts to create a way of understanding one thing by projecting onto it a view of something else. Metaphors thus highlight certain perspectives and features while masking others, especially those which are incompatible with the chosen metaphor.[101]

Metaphors can in this way keep us from seeing and understanding things differently, especially with respect to new and abstract ideas and concepts, because our experience of them and therefore our ability to see and understand them in a variety of ways is in itself limited.[101] Academic literacy is therefore critical to help us frame metaphors in ways appropriate to bridging the chasm between culture and Nature, intrinsic and extrinsic value, cause and effect, ecology and economy, and all the other aspects of environmental literacy.

Environmental Literacy

Much has already been said in this book about environmental literacy (how we think the environment works), although you may not have thought of it as such. The basic foundation of environmental literacy is understanding that change, or the principle of Creation, is a continual flow of cause-and-effect relationships that precisely fit into one another at differing scales of space and time and are constantly changing within and among those scales.

When dealing with scale, scientists have traditionally analyzed large, interactive systems in the same way that they have studied small, orderly systems, mainly because their methods of study have proven so successful. The prevailing wisdom has been that the behavior of a large, complicated system could be predicted by studying its elements separately and by analyzing its microscopic mechanisms individually— the traditional linear reductionist mechanical thinking of Western society, which views the world and all it contains through a lens of intellectual isolation. During the last few decades, however, it has become increasingly clear that many complicated systems do not yield to such traditional analysis.

Instead, large, complicated, interactive systems seem to evolve naturally to a critical state in which even a minor event starts a chain reaction that can affect any number of elements in the system and can lead to a catastrophe. Although such systems produce more minor events than catastrophic ones, chain reactions of all sizes are an integral part of the dynamics of a system. According to the theory called "self-organized criticality,[102] the mechanism that leads to minor events is the same mechanism that leads to major events. Further, such systems never reach a state of equilibrium, but rather evolve from one semi-stable state to another, which is precisely why sustainability is a moving target, not a fixed end point.

Not understanding this, however, analysts have typically blamed some rare set of circumstances—some exception to the rule—or some powerful combination of mechanisms when catastrophe strikes, again often viewed as an exception to the rule. Thus, when a tremendous earthquake shook San Francisco, geologists traced the cataclysm to an immense instability along the San Andreas fault. When the fossil record revealed the demise of the dinosaurs, paleontologists attributed their extinction to the impact of a meteorite or the eruption of a volcano.[102]

Although these theories may well be correct, such large, complicated, and dynamic systems as the Earth's crust or an ecosystem can break down under the force of a mighty blow as well as at the drop of a pin. Large, interactive systems perpetually organize themselves to a critical state in which a minor event can start a chain reaction that leads to a catastrophe, after which the system will begin organizing toward the next critical state.[102]

Another way of viewing this is to ask a question: If change is a universal constant in which nothing is static, what is a natural state? In answering this question, it becomes apparent that the balance of Nature in the classical sense (disturb Nature and Nature will return to its former state after the disturbance is removed) does not hold. For example, although the pattern of vegetation on the Earth's surface is usually perceived to be stable, particularly over the short interval of a lifetime, the landscape and its vegetation in reality exist in a perpetual state of dynamic balance—disequilibrium—with the forces that sculpted them. When these forces create novel events that are sufficiently rapid and large in scale, we perceive them as disturbances.

Perhaps the most outstanding evidence that an ecosystem is subject to constant change and disruption rather than in a static balance comes from studies of naturally occurring external factors that dislocate ecosystems. For a long time, says Dr. J.L. Meyer of the University of Georgia, we failed to consider influences outside ecosystems. Our emphasis, she said, was "on processes going on within the ecosystem" even though "what's happening [inside] is driven by what's happened outside." Ecologists, she points out, "had blinders on in thinking about external, controlling factors,"[103] such as the short- and long-term ecological factors that limit cycles.

Climate appears to be foremost among these factors. By studying the record laid down in the sediments of oceans and lakes, scientists know that climate, in the words of Dr. Margaret B. Davis of the University of Minnesota, has been "wildly fluctuating" over the last two million years, and the shape of ecosystems with it. The fluctuations take place not only from eon to eon, but also from year to year and at every scale in between. "So you can't visualize a time in equilibrium," asserts Davis. In fact, says Dr. George L. Jacobson, Jr. of the University of Maine, there is virtually no time when the overall environment stays constant for very long. "That means that the configuration of the ecosystems is always changing," creating different landscapes in a particular area through geological time.

In connection to change, Professor John Magnuson made a wonderful observation about foreseeing cause and effect.[104] He said that all of us can sense change: the growing light at sunrise, the gathering wind before a thunderstorm, or the changing seasons. Some of us can see longer-term events and remember that there was more or less snow last winter compared to other winters or that spring seemed to come early this year. It is an unusual person, however, who can sense, with any degree of precision, the changes that occur over the decades of his or her life.

At this scale of time, we tend to think of the world as static and typically underestimate the degree to which change has occurred. We are unable to sense slow changes directly and are even more limited in our abilities to interpret their relationships of cause and effect. The subtle processes that act quietly and unobtrusively over decades are therefore hidden and reside in what Magnuson calls the "invisible present."

It is the invisible present, writes Magnuson, that is the scale of time within which our responsibilities for our planet are most evident. "Within this time scale, ecosystems change during our lifetimes and the lifetimes of our children and our grandchildren."[104]

It must be noted here that while it is possible to envision such serious accidents of human misjudgment as the meltdown at the nuclear plant in Chernobyl, Ukraine, or Iraq's invasion of Kuwait, the ultimate potential destruction of the planet with respect to human life will not be as apparent. Instead, it will occur slowly and silently, like the pollution of our air, our soil, and our water—in the secrecy of the invisible present. Such an unpleasant end can be forestalled or even eliminated, however, if we humans become environmentally literate and apply the fruits of our literacy toward sustainable community development.

How might environmental literacy be applied to sustainable community development? Consider a study of the Willamette River in western Oregon,[105] the river from which my community gets most of it drinking water. The study documents an increasing incidence of deformities in fish near Portland, the largest city in the state, and shows that water running off of streets and farmland—not industrial factories—is the source of most of the pollution.

So, the question is: How can local communities clean themselves up, which they must do to be sustainable, and in so doing clean up the shared environment, giving everyone safer, cleaner air, water, and soil? Such local action can have positive global consequences for the present generation and increasingly for those of the future.

There are many novel ideas waiting to be discovered by imaginative people working to make their communities sustainable. And there are many ingenious ideas already available, such as composting vegetative materials from the family kitchen and the garden for recycling as organic material in the soil. Then there is recycling in general. With some forethought, human sewage can be disposed of by designing a system of reedbeds and wetlands, which can be operated on any scale from a household to a municipality. Nature's sewage disposal, compared to a conventional sewage treatment plant, uses little energy, requires no chemicals, looks good and natural, and can be used as habitat for wildlife.[106] In the end, we are limited only by our imagination and our willingness to risk trying something new, which brings us to economic literacy.

Economic Literacy

As with environmental literacy, much has already been said in this book about economics and its role in sustainable community development. There are, however, seven points that need to be either made or reiterated in simple terms.

First, one cannot have true economic literacy without having and applying a good working knowledge of environmental, democratic, and sustainable community literacy as well. Here, it is pertinent to repeat what John Dewey said in 1927: "A class of experts is inevitably so removed from common interests as to become a class with private interests and private knowledge, which in social matters is not knowledge at all."

Second, economics must account for the intrinsic value and long-term economic value of the functional integrity and sustainability of an ecosystem as a whole, not just the conversion potential of the pieces for immediate monetary gain.

Third, the functional integrity of an ecosystem must be considered as value added when calculating present net worth in terms of future net worth. Discounting future net worth is stealing progressively from every succeeding generation.

Fourth, it is necessary to understand, accept, evaluate, and account for the role of cause and effect beyond the marketplace in economic decisions. For example, an apparently good short-term economic decision with respect to capturing the monetary value of a renewable natural resource may prove over time to have been a bad long-term ecological decision, which, therefore, was also a bad long-term economic decision. It is thus imperative that the linkages between economics and the sustainability of the environment and human culture not only be understood but also consulted and the outcome accounted for, to the very best of our ability, prior to committing a decision to action.

Fifth, social/environmental sustainability demands that we reinvest up-front biological capital in the health of the ecosystems on which we rely for our survival, even as we reinvest economic capital in the maintenance of buildings and equipment to keep our industries functional so that they can continue turning a profit. Put a little differently, we must balance our withdrawals against our reinvestments so that we spend the biological interest of an ecosystem, but not its capital.

Sixth, a requirement of community sustainability is the achievement of the greatest possible economic independence within a well-defined bioregion through the cooperation and coordination of all communities in the region. Such economic self-sufficiency requires the conscious formation of a self-reinforcing feedback loop of local production and consumption. This feedback loop keeps local money within the local economy for as long as possible before cycling it into the bioregional economy, where it must be kept for as long as possible before letting it go into the larger world.

Seventh, all economic decisions must reconcile the necessary balance between biological capital and monetary capital within the framework of a free democratic government if a community is to be sustainable.

Democratic Literacy

Understanding the fundamental processes of a free democracy is critical if people are to see the value of their participation in making the democracy work, because, as previously stated, the principles of democracy only function when democratic processes are actually available and used. The basis of these processes allows and encourages people to give their obedience to those concepts or principles whose ethical values they hold dear and to withhold their obedience from those with which they disagree.

Consider, for example, that all we have to offer the future is options (which are choices to be made), and those options, both biological and legal, are held within the environment as a living trust, of which we are the legal caretakers or trustees for the future. Although the concept of a trustee or a trusteeship seems fairly simple, the concept of a trust is more complex because it embodies more than one connotation.

A living trust, for instance, is a present transfer of property (whether real property or personal property, such as livestock, interests in business, or other property rights), including legal title, into trust. The person who creates the trust can watch it in operation, determine whether it fully satisfies his or her expectations, and, if not, revoke or amend it.

A living trust also allows for delegating administrative authority of the trust to a professional trustee, which is desirable for those who wish to divest themselves of managerial responsibilities. The person or persons who ultimately benefit from the trust are the beneficiaries.

The environment is a "living trust" for the future. A living trust, whether in the sense of a legal document or a living entity entrusted to the present for the future, represents a dynamic process. Human beings inherited the original living trust—the environment—before legal documents were even invented. The Earth as a living organism is the living trust of which we are the trustees and for which we are all responsible.

Throughout history, administration of our responsibility for the Earth as a living trust has been progressively delegated to professional trustees in the form of elected officials. In so doing, we empower them with our trust (another connotation of the word, which means we have firm reliance, belief, or faith in the integrity, ability, and character of the elected official who is being empowered).

Such empowerment carries with it certain ethical mandates, which in themselves are the seeds of the trust in all of its senses, legal, living, and personal:

1. "We the people" are the beneficiaries and the elected officials are the trustees.
2. We have entrusted our elected officials to follow both the letter and the spirit of the law in the highest sense possible.
3. We have entrusted the care of the environment to elected officials through professional planners, foresters, and other land managers, all of whom have sworn to accept and uphold their responsibilities and to act as professional trustees in our behalf.
4. We have entrusted to these officials and professionals the livelihood and health of our environment. Through the care of these officials and professionals, it is to remain living, healthy, and capable of benefiting both present and future generations.
5. Because we entrusted the environment as a "present transfer" in the legal sense, we have the right to either revoke or amend the trust (the empowerment) if the trustees do not fulfill their mandates.
6. To revoke or amend the empowerment of our delegated trustees if they do not fulfill their mandates is both our legal right and our moral obligation as hereditary trustees of the Earth, a trusteeship from which we cannot divorce ourselves.

How might this work if we are both beneficiaries of the past and trustees of the future? To answer this question, we must first assume that our elected officials and professional people are both functional and responsible. The ultimate inheritance entrusted to the present generation for all those of the future is to pass forward as many of the existing options (the capital of the trust) as possible.

These options are forwarded to the next generation (in which each individual is a beneficiary who becomes a trustee) to protect and pass forward in turn to yet the next generation (the beneficiaries who become the trustees) and so on. In this way, the maximum array of biological and cultural options can be passed forward in perpetuity—the essence of sustainability.

If, however, the elected officials and professionals do not fulfill their obligations as trustees to our satisfaction, then their behavior can be critiqued through the electoral process and/or the judicial system, assuming that the judicial system is both functional and responsible. The invisible present embodied in our decisions as trustees of today can then create a brighter, more sustainable vision for the generations to come, who are the beneficiaries of the future when they stand in their today.

In order for this to happen, however, we must actively participate in the democratic rule of our communities so that they become as sustainable as possible in partnership with their surrounding environments within a bioregion. We must understand and accept that a sustainable community is, in a sense, the institution in which the living trust is housed and protected. We must also make our judicial system just and responsible to all generations, something we have not yet chosen to do. And it is, after all, only a choice, which is the very foundation of democracy.

Democracy sets up and maintains the information/political feedback loops through which human values, intuition, information, and cultural innovations are funneled into the melting pot of community literacy.

Community Literacy

It is in a freely democratic community that the cultural gold is separated from the dross. The gold is a community's potential to behave like an intelligent, moral, innovative, and freely democratic organization while on the road to sustainability and beyond. The first step along the way is for a community to identify itself as a community, or as John Dewey said in 1927: "Unless local communal life can be restored, the public cannot adequately resolve its most urgent problem: to find and identify itself."

Identity

Who are we now, today? This is a difficult but necessary question for people to deal with if they want to create a vision for the future. The vision they create will be determined first by how they identify themselves as a culture and second by how they identify themselves as a civic organization, which in turn is defined by its governance. The concept of citizen government means that "citizens must possess the skills and dispositions to act as rulers and know when this obligation is required."[107] In other words, citizens must know when the good of the community, present and future, is at stake.

Therefore, the self-held concept of who a people are culturally and how well their community governance represents them is critical to the sustainability of their future. Their self-image is crucial because it will determine what their community will become socially, which in turn will determine what their children will become socially.

A major problem facing communities today is that people are no longer thought of as people but rather as a group of "publics," which is an amorphous aggregate of individuals and their preferences. In this sense, "public" means whatever aggregate of individuals is being measured at the moment, such as the public as market player, as skier, logger, cattle rancher, consumer, scientist, and so on. But in none of these perspectives is the public thought of as whole persons whose humanity supersedes whatever else they might be.[107]

Thus, how well a community's core values are encompassed in a vision depends first on how well the people understand themselves as a culture, second on how well that understanding is reflected in their self-governance, and third on how clearly it is committed to paper. Only after people have dealt with who they are, today, can they determine what legacy they want to leave for their children and create a vision with which to accomplish it, because only then do they know.

Visions will vary greatly, depending on how a community is defined. A rural community, for example, will include its immediate landscape and perhaps even its relationship to neighboring communities and the bioregion. Within the "inner city," however, a community may be one square city block and its relationship to the four neighboring blocks facing it. Regardless of how people define their particular community, their success in self-governance depends on their sense of citizenship, which is currently entangled with the term "public."

Citizenship

The many schizophrenic splits in the concept of the term "public" is the result of what is missing, according to Manfred Stanley.

> The broad core of the classical approach to citizenship stresses a shared constitution that embodies not only rules but a founding myth, a sense of collective moral history, a common-law tradition, and some conception of a good way of life. The principle of public integration is "civic friendship," a concept designed to call attention to bounds transcending mere commercial or military utility. In this sense of public, policies are designed to affect the fortunes of the commons in a particular, historically contingent, moment of its moral development. This is to say, public policies are initiated and evaluated in light of a mythic vision of…a good society.[107]

Citizenship is not some abstract quality of an isolated individual action. Rather, citizenship is the cumulative effect of a sequence of actions in relationship, where the whole meaning arises from what might be called their ritual practice. A ritual refers to conscious events that come in regular sequences and acquire meaning from their relations to other events in the sequence.[108]

Outside of their full sequence, the individual elements become lost, their existence imperceptible and nonsensical, and their meaning cloudy. Consider, for example, that the meaning of each day of the week comes from its sequential relationship to all the other days, but only so long as they are in the correct sequence. Mix them up and they lose their meaning.[108]

Citizenship, which is based on mutual civility, recognizes that it is the quality of human relationships that either allows and fosters the sustainability of a community—or kills it. Vinoba Bhave has some clear thoughts on civility as the basis of citizenship:

> I am moved by love. I do not deal in opinions, but only in thought, in which there can be give and take. Thought is not walled in or tied down; it can be shared with people of goodwill; we can take their ideas and offer them ours, and in this way thought grows and spreads. This has always been my experience.…It is open to everyone to explain

their ideas to me...and they are free to make my ideas their own in the same way.

There is nothing so powerful as love and thought; no institution, no government, no "ism," no scripture, no weapon. Love and thought are the only sources of power.[109]

As important as civility and citizenship are, however, sustainable community development is possible only to the extent that people keep learning. One innovative way of learning in a democratic setting is study circles.[110] A study circle is a small-group discussion format to seek understanding and common ground when people face difficult issues and hard choices. Study circles reflect a growing conviction that collective wisdom resides in groups, that education and understanding go hand in hand, and that learning can truly be available for all.

The circular shape of the study group is important and has its roots in antiquity. In medieval literature, for example, brave knights came from across the land to be considered for membership at the Round Table. King Arthur designed its circular shape to democratically arrange the knights and give each an equal position. When a knight was granted membership at the Round Table, he was guaranteed equal stature with everyone else at the table and a right to be heard with equal voice.

In study circles, participants learn to listen to one another's ideas as different experiences of reality rather than points of debate. Although they may not agree, they learn to accept that, just like blind people feeling the different parts of an elephant, each person is limited by her of his own perspective, which is derived from his or her own experiences in life.

By managing the process themselves, participants engage in the practice of democracy. In a study circle, there is equality, respect for others, and excitement about exchanging ideas. This environment is ideal for people to practice the most fundamental aspects of democracy by reaching conclusions or making decisions through talking, listening, and understanding—through sharing.[110]

Sharing is the central connection in study circles. Participants are encouraged to act as whole people in that they are not required to separate feelings, values, and/or intuition from intellectual thoughts concerning any topic. They are allowed to think systemically as opposed to being placed in a straitjacket of intellectual isolation.

This sharing as whole individuals allows each person to assume the role of teacher, student, leader, and follower at different times in the study circle, which is critical to the viability of both the democratic process and sustainable community development. Because no one person possesses, with equal skill, all of the talents necessary for the practice of either democracy or sustainable community development, it is vital that individuals learn to accept and share the many facets of their personalities to the best of their ability.

People seldom partake of study circles just to learn the so-called objective facts; rather, a study circle deals with real daily problems in the lives of the participants and is thus education in and for life. It is imperative that what the participants learn is grounded in their own experiences and in the real problems and issues they face daily.[110]

Study circles bring people together to talk and to listen; to act and feel as if they are part of a community; and to practice equality, acceptance of ideas and diversity of people and points of view, democracy, and the connectedness of sharing. If an increasing number of people became involved in study circles, it might become clear that the apparent apathy that Americans exhibit toward education and participation in politics is really a disguise for a deep hunger to learn within the safety and nurturance of community.[110]

I say this because, as Myles Horton expressed it: "The fact is that people have within themselves the seeds of greatness, if they're developed. It's not a matter of trying to fill up people, but to fulfill people." And I believe this with every fiber of my being! "This is all good and well," you might say, "but how can we fulfill people?" I think the answer lies in helping communities become intelligent, moral organizations with a vision of social/environmental sustainability to be enjoyed in the present and passed forward as an unconditional gift for the generations of the future.

Community as an Intelligent, Moral Organization

Our American society has a peculiar split personality with respect to the issue of centralized power. We openly criticize the government-controlled economies and news agencies of other nations because we believe that centralized power leads to the suppression of liberty and bureaucratic corruption and waste. But most of our government bureaucracies and private institutions operate under the assumption that internal centralization of power leads to efficiency and that free choice creates and maintains inefficiency.[111]

Efficient is defined in the *American Heritage Dictionary* as acting or producing effectively with a minimum of waste, expense, or unnecessary effort. Efficiency is defined in the same dictionary as the ratio of the effective or useful output to the total input in any system; especially, the ratio of the energy delivered by a machine to the energy supplied for its operation. Because individual rights and the freedoms they protect invite inefficiency, power is centralized under the guise of efficiency to omit the human dimension whenever possible and, with it, the democratic process.

The question thus becomes: How do we as a society build democracy back into our communities? To examine this question, I will borrow heavily in the following discussion from some articles and a book by a farsighted husband and wife team, Gifford and Elizabeth Pinchot, whom I acknowledge with gratitude.[111,112] While the Pinchots deal with corporations as intelligent organizations, I have adapted their work to local communities as intelligent organizations. We begin by discussing the distribution of power.

Balancing Power

Because absolute power corrupts absolutely and effectively squanders the human spirit and kills the human will, every revolution has the same theme—balancing the power by flattening the pyramid of the controlling hierarchy. To keep our governing bureaucracies flat, especially those that affect communities, and to make the most of the flattened structure, we must have tools that reach far beyond dehumanizing efficiency. For communities struggling toward social/environmental sustainability, these tools include empowering education for all and the rights of free speech, of assembly, of democratic decision making, and of joint and private ownership, but with moral provisions to protect the ecological and personal rights of the generations of the future.

Many revolutions have a common denominator, which is a basic shift of the day-to-day control and feedback systems from the enthroned hierarchy of power brokers to committees of people. The current revolution of quality (as opposed to quantity) is the most successful attack on bureaucracy, and it has inspired much progress in developing systems of work, including communities, that are both democratic and collaborative.

In a community, it begins with the recognition and acceptance that the quality of intelligence needed to meet the requirements of social/environmental sustainability demands the expertise and creativity of every person. This means teaching members of the community the skills necessary to ask relevant questions, seek answers, make democratic decisions, and then implement them.

As soon as the common citizens of a community empower themselves to make decisions in interdisciplinary committees, the power invested in the controlling hierarchy is challenged and the belief in the efficiency of specialization is undermined because the teamwork of a dedicated committee remains focused on and acts on the issue before it. What makes this work is the ethical underpinnings of the committee's other-centeredness, which directly contradicts the self-centeredness of the controlling hierarchy.

The success of a sustainable community as an intelligent organization depends on its sense of and commitment to an ethical foundation for its day-to-day operations. Ethics is no longer a luxury, because in these times of growing limits of renewable natural resources on a global basis, ethics is the necessary staple in any sustainable human endeavor.

There are no more plentiful, cheap resources to solve our problems, and continual growth as a panacea no longer works. We are now faced with a suite of dilemmas: Do we need more freedom or more control? Do we need more soft humanism and equality or more hard-line discipline and sacrifice? More collective cooperation and coordination or more individualistic risk and initiative?

The bureaucratic solution of more control (more hard-line discipline and sacrifice, more individualistic risk and initiative, which then needs to be controlled by the hierarchy) no longer works. We need more open, self-organizing communities, modeled on Nature, which have the capacity for self-renewal in a way that is more than simple self-preservation.

The inner security encompassed in the capacity for self-renewal is why old-fashioned values are having a renewed following. We are rediscovering (relearning) that we only bring our best selves to the table when each person is valued for what he or she has become as an inner person, for his or her contributive talents, and when there is sufficient freedom and safety to share the personal gift he or she has to offer. This in turn is either nurtured and protected by the ethics of individual conduct in the collective of a local community or it is destroyed.

Ethics

Freedom, in turn, is real only when it is exercised with a state of mind well beyond the limits of self-centered individualism. A self-organizing human community depends on each person within the community behaving in an ethical manner, even when no one is watching and there are no penalties for misbehavior. For trust to grow and manifest itself as cooperation, coordination, and adaptability, that trust must be based first and foremost on faith and second on an assurance of the goodness of others within the community.

For a community to be sustainable, it must be organized in such a way that freedom and personal initiative can cohabit with cooperation, coordination, compassion, and a highly integrated harmony. For a community to be sustainable, it must be grounded in the ethical basics of freedom and democracy, which include recognizing, understanding, and accepting the value of the following: (1) diversity; (2) decentralized power; (3) shared and revolving leadership; (4) continuous self-testing; (5) local, regional, national, and global ethics; (6) acting for future generations; (7) making sure that jobs and economic opportunities are available for young people; and (8) practicing the Golden Rule.

1. Diversity—For freedom to be real and collective, it must encompass, value, nourish, and protect diversity. This means that the uniqueness of each and every person, expressed collectively as diversity, is inherently of equal value, is an open-ended potential for contribution, and is equally necessary to the sustainability of the whole.

Within each ethnic and cultural group there resides an enormous diversity that individuals bring into a community—diverse experiences and therefore frames of reference, styles of organization, talents, and competencies. The freedom not only to be diverse but also to be valued for it is the leavening of dignity and effectiveness within a

democracy. Assigning equal value to each of us brings out our uniqueness and drives the success of interdisciplinary committees, which are
the foundation stones of sustainable community development.

2. Decentralized Power—For a community to be flexible, responsive,
and adaptable, intelligence and creativity must be distributed throughout it, which means that each and every person must be encouraged
to use his or her mind and ingenuity and must be openly valued for
doing so. To be effective, people must interact in such a way that new
ideas and information are shared and used continuously and rapidly.
It is important to remember that there are only good ideas, some of
which will not work, but there are no bad ideas. To label an idea as
bad steals dignity from the person offering it and may cause that
person to withdraw.

To achieve such flexible, responsive, and adaptable intelligence
and creativity, the power to decide and act must be distributed throughout the community. The forces for maintaining centralized power in
practice are so great that for a community to be effective, it must raise
the sharing of power to an ethical principle, backed up with some
level of control through recognized authority.

In practice, the most viable communities are promoting the emergence of new, informal, interdisciplinary committees to develop as
necessary across all traditional boundaries within and outside of the
community in a concerted effort for the community to become sustainable. This activity replaces the simplistic hierarchy with an amorphous
and fluctuating complexity of ever-adjusting relationships.

When the members of a community are increasingly self-empowered and self-organized, simple rules and rigid policies are insufficient
to guide them. There must be a shared vision of where they want to
go and what they want to accomplish embedded in a deep respect for
the ethical foundation on which they stand, because the ends, however nobly perceived, never justify unethical means. The authenticity
of leadership is providing ethical and effective power to the people.

3. Shared and Revolving Leadership—Authentic leadership must be as
widely shared within a community as are intelligence and creativity.
Free people can only cooperate and coordinate with intelligence if
they share a vision toward which to build, in addition to which they

must have the ability to test and to renew themselves with accurate feedback. People find it easiest to trust one another when they are working actively toward a shared vision based on a common and widespread moral ethic.

Shared leadership integrates the paradoxes inherent in human relationships. The new organizational designs will encourage equal measures of freedom, cooperation, and coordination with equal doses of fiscal discipline. Like all paradoxes, the opposites must by unified to be whole and benign. Anyone who ignores one side of a paradox will be burned by it. We must therefore develop in ourselves both the analytical and the intuitive, the shrewd trader and the compassionate compatriot, the entrepreneur and the communitarian, the individualist and the egalitarian, the competitor and the partner. Wholeness, which requires constant testing, is the hub around which sustainability revolves.

4. Continuous Self-Testing—Communities that are energized with self-organizing committees and projects, that are flexible and responsive, gain needed cooperation and coordination from continuous self-testing against broad ethical principles and a shared vision. To give people freedom and power is one thing; to assure their ability to use it wisely in quite another.

To examine the wisdom with which members of a community, and therefore the community itself, use their freedom and power requires self-testing against the shared vision and the ethics of its foundation. Such testing not only must occur continuously throughout the entire community but also must promote truth and free and unlimited access to information.

5. Ethics on a Global Scale—Ethics must be like ripples in a pond—ever expanding. While freedom and power extended to people imbued with a deep sense of local community must initially inspire the ethics of community, ethics must grow beyond the parochial definition of what the individual is initially part of. In other words, ethics must ultimately grow beyond a local community if the community is to become a free, empowered, and responsible member of the bioregion, nation, and world.

When a local community extends it ethical sphere beyond itself, it achieves a kind of aliveness that is self-reinforcing and contagious. The community is teaching by example.

6. Acting for Future Generations—We must understand, accept, and be accountable for the circumstances of the future because they will always be rooted in our decisions and behaviors in the present. The most difficult and important challenge of our times, therefore, is to discover and create ways in which social/environmental sustainability can be achieved. To accomplish this, we must work diligently and honestly to determine, as best we can, both the cultural capacity and carrying capacity of each local community, county, state, bioregion, nation, and ultimately the world.

We can only accomplish this goal by protecting the options for the generations of the future and by being mindful of how we behave today. Put simply, our choices and actions set in motion forces that create the circumstances the children of tomorrow will inherit as the legacy of consequences we have left for them. The question we must ask ourselves is: Would we want to—or could we even—contend viably with the circumstances that we are leaving for the generations of the twenty-first century, such as the job prospects and economic opportunities for young people?

7. Making Sure that Jobs and Economic Opportunities Are Available for Young People—The most difficult and important social/technological challenge of our time is to find ways to bring the whole human population to a quality of living that nurtures human welfare and honors human dignity while simultaneously protecting the sustainability of the environment. This cannot be done with existing technology, much of which is not only environmentally dangerous but also contrived to eliminate people's jobs.

Those corporations that value and nurture their people, because they see in them the real corporate wealth, will have the competitive edge in the future. The competitive edge in this case is the loyalty and imagination that each person brings to his or her job. The competitive edge includes sound ecological principles applied to industrial goals and practices not only because they will save money in the face of inevitable regulations and taxes to come but also because good public relations is good marketing.

Wise corporate management will therefore ensure that young people have quality jobs and sound economic opportunities if they want to maintain their competitive edge in the future. Downsizing is anti-competitive over time because it squanders a corporation's real wealth—the loyalty, imagination, and the healthy long-term interpersonal relationships of its people, which brings us to the Golden Rule.

8. Practicing the Golden Rule—"Do unto others as you would have others do unto you." This statement not only seems simple enough but also is a cardinal rule of local community life and survival. Any community that is sustainable has achieved its harmony by treating its own members, as well as it neighbors, as equals, with compassion, dignity, and respect.

The Golden Rule is needed internally in the local community to assure peaceful discourse, freedom and empowerment, cooperation and coordination freely given, and a sense of unity through mutual compassion. The Golden Rule is the ethical pivot of life's compass on which the behavioral needle swings in seeking its direction, and the direction chosen determines the lasting value of that which is done.

These days, it seems that what we call the "realities of life" are harsh indeed and getting harsher. This harshness can be softened, however, when we know we can count on others and they on us, which is the underlying value of a local community.

In some deep inner recess, everyone knows that a renewed commitment to personal ethics; a shared vision; and farsighted, other-centered behavior are necessary to revive the real, personal sense of local community. Many of today's problems result from moral lapses in self-discipline and personal accountability for one's own behavior, as well as materialistic, self-centered overindulgence. To prevent these problems from becoming an untenable burden for the generations of the future, we must all have an unshakable desire and commitment to discern that which is truly ethical and apply it both within ourselves in secret and among others in public.

To do this, we must find people who are content to be servant leaders, people who can help their followers instill within themselves the ethical awareness necessary to be other-centered by sharing their creativity, intelligence, and competency for the sustainable good of their local community. Such leaders must foster intelligence and honor intuition among members of the community, especially those who are willing to serve as volunteers on local committees.

Intelligence and Intuition

How can the intelligence and creativity of every member in a community be liberated? How can the ideas, inspiration or intuition, and intellectual analysis of members be integrated into both rapid decision and action for the sustainable good of the local community?

No centrally conceived design is capable of producing or allowing the freedom necessary to empower individual intelligence and honor individual intuition while simultaneously knitting together free thinkers in coordinated action with a single focus. Organizations that simultaneously foster intelligence and honor intuition must grow out of the convergence of processes within a community. For this to happen, however, seven conditions must be established based on freedom of choice and democratic participation: (1) widespread truth, evaluation, and rights; (2) liberated employees and committees; (3) freedom to be enterprising; (4) justice and equality; (5) processes for self-management; (6) voluntary networking; and (7) limited community government.

1. Widespread Truth, Evaluation, and Rights—People can only make responsible choices if they know what is going on. Whereas self-centered bureaucrats tend to hoard information, thinking it their source of personal power, a community must create a free exchange of information if it is to act with intelligence.

The Central Intelligence Agency (CIA), for example, is the antithesis of an intelligent organization. It is instead a body designed to collect information in secret, analyze it in secret, decide in secret who should know what, and parcel the information out piecemeal in such a way that the agency's secrecy is not only retained as its source of power but also constantly growing.

If, therefore, a local community is going to act as an intelligent organization, it must make available full financial information and train all community employees, and any member of the community who wishes to know, how to read the financial statements. All activities that in any way affect the common good of the community must be regularly evaluated and those evaluations immediately posted for all who are interested to read. It must be safe to openly discuss problems that arise and any strategic options to resolve them. There must be a constant iteration and reiteration of how each part fits into the whole and how the whole requires and cares for each part. Finally, freedom of speech and freedom of the press are essential if a community is going to act with intelligence, as is the right of inquiry and of learning for the betterment of the whole, present and future.

2. Liberated Employees and Committees—Behind nearly every recent innovation in sustainable community development is the superior effectiveness of team-oriented employees and team-oriented committees. Team-oriented committees are the basic building blocks of a community as an intelligent organization.

To reap the benefits of autonomous, empowered committees, we will need the following: (1) a committee must have meaningful authority; (2) it must have free choice of its task, partners, members, and connections; (3) each committee must be evaluated and rewarded as a whole; (4) the people must be trained in the processes of self-management and systems thinking; (5) cooperation and coordination must come from within the committees themselves rather than from a supervisory level downward; and (6) the committee's purpose must be worthwhile in and of itself and must be integrated within a larger vision, the achievement of which is of value to the members of the committee as a whole.

3. Freedom to Be Enterprising—Intelligent communities release the creative energy of individuals and committees both by making it safe to be creative and by preventing monopolies of power from squelching them.

4. Justice and Equality—The kind of democracy needed for members to design and redesign their own community must make creativity and intuition both welcome and safe and must encourage their expression through direct participation in a free democratic process. Liberated members of a local community must be trusted to have the sustainable good of the community at heart.

Communities designed to bring out the responsibility, intelligence, and creativity of every member will depend on internal systems for guaranteeing justice and equality. These internal systems must protect the people from imbalances in power and use sound processes of transformative facilitation to resolve conflicts.

Communities as intelligent organizations will grant freedom that is limited by clearly stated internal laws and an effective system of justice, rather than depending on bureaucratic supervision to prevent abuses of power. The results are better control and, within that, more freedom to be creative.

5. Processes for Self-Management—Before people are likely to become involved in a self-managing committee, they must know that there are processes in place to support them. When these processes are effective, they engage people in collaboratively managing the whole, which means that all voices are heard and respected so that innovation can take place. This results in the implementation of more new ideas and fewer unthinking mistakes.

Intelligent communities involve all employees and members of volunteer committees in creating the larger context for their work. Consent and consensus guide, as much as possible, the design of the policies and institutions necessary to steer the community toward fulfilling its vision. The result is systems, strategies, and policies that respect the needs of people to feel productive and to continually adapt to changing conditions by experimenting with new ideas.

6. Voluntary Networking—For a community to be flexible and responsive to changing conditions, intelligence and intuitive insights must flow from every involved member of the community. Every person must interact in such a way that new information is rapidly disseminated and applied. Only voluntary networking can forge the linkages necessary for such massive, flexible interconnected conduits of information, because they must be created moment by moment through the choices people make in establishing the connections they need to accomplish their respective tasks.

7. Limited Community Government—For any society (including a local community) to exist, it must have some form of government to ensure personal rights, safety, and other basic requirements of the common good. The central government of an intelligent community is therefore limited in scope and power because its primary role is to create and protect conditions that allow self-empowered members of the community, both employees and volunteers alike, to build the systems necessary to guide the community toward its shared vision.

Status

One of the tasks of leadership is to raise people above selfish, parochial concerns by inspiring them to work together by example for the common good of the whole. The hallmark of good leadership is a community so focused on its shared vision that the members have little desire or energy to squabble over status. A leader can help people satisfy their need for recognition in at least four productive ways: (1) make belonging to the group a source of pride, (2) recognize inner growth and self-control, (3) spread honest recognition throughout the group, and (4) move beyond the postures of dominance and submission.

1. Make Belonging to the Group a Source of Pride—Leaders inspire their followership by focusing on the shared vision with such authentic enthusiasm that people *want* to join in the adventure of achievement. They lead the group in a constant celebration of its being and its moral growth, as well as it material achievements. A leader sets both ethical and material standards of structure and excellence by example, so that every member can see a path to achievement. By helping the group see itself in the light of its own collective uniqueness and skills of both revolving followership and leadership, the issue of relative status within the group is diffused.

2. Recognize Inner Growth and Self-Control—By rewarding people for the most difficult thing any of us ever do, which is to grow through self-discipline beyond where we feel comfortable, people are valued for what they are and are becoming in the best sense of themselves. What greater value is there for a person than the inner growth that leads to a greater sense of consciousness and the personal freedom that comes with it? What greater value is there for a community than to have as its member a person of such courage?

3. Spread Honest Recognition Throughout the Group—Sensitive leaders want all the people they serve to receive recognition when it is justly earned, for only then does it have real value. Recognition or status is variable within any system, but it can be generally increased by seeing that the potential of people is released in settings where each individual is valued equally as a person first and foremost. Beyond that, individual recognition is forthcoming based on how a person grows and serves the community as a whole.

4. Move Beyond the Postures of Dominance and Submission—One of the ways we humans read status is by noting dominance and submission in others and feeling it in our own interrelationships, just as other social animals do, such as monkeys and wolves. Centralized governmental power within communities taps into relationships based on dominance and submissiveness as a matter of control, which means a perpetual struggle for dominance within community government, where the "successful" ride fear to the top by making those around them submissive. In turn, others ride behind the truly fearsome by identifying with reflected fear to intimidate those of lower rank.

Such behavior is grievously flawed, however, as a basis of an intelligent community because those on top are isolated from what is going on beneath them. Isolation leads them to believe their own self-serving version of reality, and no one dares contradict them. The resulting forced submission brings out resentment and its accompanying resistance, which destroys creativity, initiative, self-esteem, and sound choice. There are, however, some ways around the either/or relationship of dominance versus submission:

1. To become fully functional adults, we must each find our own place in relationships that are beyond both dominance and submission; for example, the friendship of equals, the working partnership of equals, the temporary relationship of guest and host, none of which are based on dominance and submission.
2. By having the courage to design human systems that lead toward relationships of equals, a community leadership can encourage people to give up their submissive behavior and accept the responsibility of self-guided action.
3. Move toward the revolving status of required expertise, which necessitates giving up the status of a "pecking order." Revolving status of expertise means that when a given person's expertise is required, he or she automatically assumes temporary leadership until someone else's expertise is needed, in which case she or he assumes temporary leadership, and so on. In revolving status, the unspoken rule is that I am dominant in my area of expertise based on my greater knowledge and you are

dominant in yours based on your greater knowledge; I will recognize your dominance and you in turn will recognize mine.

With time and personal growth, it even becomes possible to move beyond the need for the status of expertise as people move naturally into the next step of interdependence, which brings more activity into the mutual space between the territories of expertise. This is the domain of equality and intelligent interaction from which partnerships are formed.

Partnership is a relationship that is beyond the concerns of relative status. In partnership, the issue is the smoothest function of the whole for the common good, where each person cares for the other as an end, not as a means. By faithfully looking out for one another's dignity and interests, the partners build a mutual trust that allows both cooperation and coordination, as well as sharing resources.

4. One advantage of a free democracy is that it allows the growth of a self-organizing human system constructed on the freedom of choice, which expresses itself in the voluntary associations of people from all walks of life who share a common vision toward which they are building. A system of many choices works because a decentralized democratic system of government, for all its faults, is more egalitarian than is a centralized hierarchy as a system of government. It provides a firmer foundation for relationships based on choice rather than one founded on compulsion.

5. Leaders of intelligent communities tend to select symbols that blend in and make them accessible, such as common parking lots and modest dress. The intelligence of a community can be increased by reducing the differences in status and thereby focusing people's efforts on the community's shared vision, rather than on gaining power over others.

6. Create an array of sincere rewards for many winners, which means focusing attention on areas were everyone can win, as opposed to salaries.

7. Give recognition to those who use status wisely, so the status they achieve can be replicated. Pay close attention to which behaviors are being rewarded and see that good, morally honest behaviors win.

8. Treat everyone with dignity and respect. People with a good sense of dignity and self-respect are less influenced by how others view them and thus are less vulnerable to the blandishments of any system that allocates external status. Wise leaders conserve, create, and apportion recognition with the same care they devote to ethical conduct and responsible fiscal management. They build on voluntary agreements among equals within self-organizing human systems, which brings us to the notion of moving beyond bureaucracy.

Beyond Bureaucracy

When decisions are made from above and passed down, they affect things that people at lower levels in the hierarchy know about, but these people are neither asked for their information nor listened to when they offer it. They therefore stand idly by watching their community do foolish and wasteful things. They become inured to the decision makers' lack of focus in the relatively small daily decisions, which cumulatively affect the larger decisions the decision makers think they see with such clarity.

The way out of this closed box is to install leadership that will replace blaming and quick fixes with a focus on discovering the dynamic relationships within the community and then open the community to changes, which permit the widespread use of intelligence, intuition, and creativity. This means that the community must begin to live on a free and open exchange of information, including free speech and free press, and anything blocking the flow must be viewed with suspicion.

Truth must be a community's primary value, so that no thought will be silenced or punished. Everyone must be free to comment; any manager who attempts to silence unfavorable comments *must be removed.*

Intelligent communities have many layers of self-organizing human systems, each of which is interrelated to the others. Consider, for example, the human body, in which the intelligence to form a cell is found within each and every cell, not in the brain. And even in the brain, intelligence is an emergent property of the interaction among the cells of the brain, and not the sole possession of any particular cell. Self-organizing communities, like the brain, require too many subtle connections for any one person or committee to design. The most intelligent communities are designed moment to moment by people interacting freely to get done the things they need to do.

Whenever feasible, an intelligent community opts for the freedom of choice, as opposed to a monopoly of power. Choice may appear to be less efficient than a monopoly in the short term, but over time freedom of choice leads to self-motivation and continual innovation, which increases effectiveness as well as cost efficiency. It also allows people to make mistakes together, live with the consequences, and clean up the mess without recrimination from above.

The best communities are therefore designed to give nearly everyone control over something so that everyone has sufficient choices about themselves that they take charge of their own direction and effectiveness. While working in Nepal, for example, I had a thief amongst my employees. After finding out who it was, I officially placed the thief in charge of all of my laboratory equipment. I never lost another piece of equipment to theft, and the thief was exceedingly proud of that.

An intelligent community is, in effect, a community of free people who feel safe enough in both their community and their own competence to form groups through which they cooperate and coordinate with a minimum of egotistical turf wars. This is possible because when a given person's expertise is required, he or she automatically assumes temporary leadership until someone else's expertise is needed and she or he assumes temporary leadership, and so on. The paradox is that to get people beyond defending turf, they must be allowed to have turf within acceptable social constraints.

Intelligent communities can only be built with trust and the freedom trust endows, which means behavior must be ethical at all times. To focus on ethical behavior is to focus on a worthwhile purpose. The higher the purpose, the greater the ethical integrity and the greater the community's intelligence. Once a community's ethical integrity and intelligence are raised in proportion to the quality of its purpose, all incongruous systems within the community, meaning those based on lower-level purposes, will be replaced automatically and rapidly, because within the domain of purpose, higher ones tend to drive out lower ones.

Just as domesticated dogs behave like puppies all their lives when compared to wolves, so too many of us have sold our dignity as adults for the steady wage. Therefore, communities that learn and apply the secrets of freedom, voluntary interdependence, and the co-responsibility of community will attain and continually develop intelligence, intuition, and creativity.

The exemplary symbol for the beginning of an intelligent organization is the lever above the head of every assembly-line employee at Toyota. This lever gives the employees the power to stop the assembly line if something is wrong. Inherent in this lever is a great leap of faith, which is the beginning of a revolutionary shift from trusting in the power of bosses to trusting in a well-designed human system with built-in freedom that honors the integrity and intelligence of the ordinary worker.

Freedom and meaningful communication are difficult to find in our current human world. Any community that incorporates more of either will reap accruing dividends and help to reshape communities in the next century. But in addition to the human component, there is also the sense of place inspired by and through the feeling of one's surrounding environment in relationship to one's local community.

Community and a Sense of Place

In speaking of community as a sense of place, I do not in this case mean the physical location in which a particular community happens to rest, but rather the human/spiritual component of community as expressed most clearly in the monastic concept of *stabilitas* or stability.[113] Simply put, stability in community implies not only abiding in a particular place but also identifying oneself with the community in all its works, its ups and downs, its tensions, joy, and sorrows. Stability in this case means perseverance in and with a community over the long term for its common good.

In a monastic sense, the vow of stability is based on the fact that a monk, under the appearance of greater potential, which we might call a "greener pasture," may follow a path from one community to another, and in so doing lose the good already at hand. The purpose of the vow is to make a monk (man or woman) realize that stability in and of itself is an immense good and that in a vast majority of cases constitutes a much greater good than might be gleaned by changing communities.

If a monk will retain stability, he or she will be able to effect the greatest and most important change—that of oneself, the inner transformation into a more conscious human being open to the balance between spirituality and materiality. If we continually seek outer distraction by moving from community to community, we will continually lack the focus to achieve the inner transformation that could be ours here and now in this community.

In monastic life, it is not sufficient to remain in the same place. It is also necessary to be under the direction of a spiritual teacher. In a secular community, this might be translated to mean that it is not sufficient to remain in the same place; it is also necessary to be committed to a high communal purpose. Thus comes stability in commitment.

Community is for the people who make it up. Stability, therefore, is not primarily for the perfecting of the community in the sense of making it more stable and orderly. Rather, stability is to root the people in the search for a common humanity in which such things as love, faith, trust, mercy, compassion, sharing, and justice become the human foundation of democratic governance.

By committing oneself to become a member of a community, one is above all making a promise of stability, in which we take upon ourselves the responsibility of offering our help to other members of our chosen community, both in the material and in the spiritual. The commitment of stability is not only a promise to persevere within the physical community but also, and most importantly, a promise to persevere in our commitment to live as best we can the love, faith, trust, mercy, compassion, sharing, and justice that make our community a place of safety and personal value.

A community may be a safe place, but to hold people within it and to bring people back after they have experienced more of the world, a community must have things to offer that make staying and/or returning a desirable option. In large part, sustainable community development is creating just such value.

I noticed in Malaysia, for example, that some of the villages did not have many children visible in them. When I asked the reason, I was told that the government sent them away to school because the village had none. But in sending the children away to school, their sense of their home community will diminish and its value will be "educated" out of them. To instill a sense of value in the children for their own community, the government would be wise to secure a teacher or teachers for the village.

Having a school in the village means there is a better chance that some of the children will stay and add to the labor pool necessary to internal sustainability. It also means there is a better chance that those who do leave will come back, bringing with them ideas and skills necessary for the betterment of the community as a whole. This is the hero or heroine's journey described by mythologist Joseph Campbell in *The Hero with a Thousand Faces.*[114]

In this journey, a youth travels away from his or her village to explore the world. Having found adulthood, expressed as some sense of self-mastery, the man or woman returns home and teaches the villagers through service what he or she has learned. But there must be a compelling reason for youth to return as adults.

The higher the moral purpose of a community's vision, the greater its commitment to sustainable development, the more of an intelligent organization it becomes, and the more value the community has to offer those young members who will in time become its leaders. Thus, the more personal the value one feels about one's community, the easier it is to identify one's community with a sense of place, which is, after all, the foundation of sustainability community development.

REKINDLING THE SPIRIT OF COMMUNITY

Rekindling the spirit of community is a choice—no more, no less. The same is true of sustainable community development. Whatever we choose to think about, whatever we choose to focus on, that is what we create. As anthropologist Margaret Mead says: "Never doubt that a small group of thoughtful, committed citizens can change the world; indeed it is the only thing that ever has."

Therefore, "Whatsoever things are true, whatsoever things are honest, whatsoever things are just, whatsoever things are pure, whatsoever things are lovely, whatsoever things are of good report...think on these things."[115] Thinking on "these things" reminds me of the life cycle of a salmon, which epitomizes the destination of choice.

A reddish orange egg is deposited in a redd (the gravely stream bottom that serves as a nursery for salmon) in the headwaters of a

Pacific Coast stream. There the egg lies for a time as the salmon develops inside. In time, the baby salmon hatches and struggles out of the gravel into the open water of protected places in the stream. There it grows until it is time to leave the stream and venture into life. It can go only one way—downstream to larger and larger streams and rivers until at last it reaches the ocean.

After some years at sea, the inner urge of its species drives the adult salmon along the Pacific Coast to find the precise river it had descended years earlier. It must make a critical decision. If it selects an incorrect river, it will not reach its destination, regardless of all the other choices it makes. If it swims into the same river it had descended, it is on the correct course—until it comes to the first fork and must choose again.

Regardless of its immediate choice, however, in the lower reaches of the drainage basin, the water is deep, polluted, and relatively warm, its current placid. Here the salmon swims easily in the wide river, where there is much to distract it from its upstream journey.

Each time the salmon comes to a fork in its journey, it must make a choice and must accept what the chosen fork has to offer and forego the possibilities in the one not taken. But each time it chooses the correct fork, the salmon finds the water confined within an ever-narrowing channel, flowing progressively swifter, cleaner, clearer, and colder than from where it just came.

As the streams' banks become more confining, the salmon finds its focus on its destination becoming sharper and more urgent. Now the obstacles in the streambed, such as large boulders and swift waterfalls, are as nothing, so focused has the salmon become, so clear in its determination, so urgent in its inner need to arrive at the particular spot within the designated time. When the salmon reaches this state, its focus is so concentrated that it finds the current's force diminished against the internal power of its life's spirit, its inner drive to reach its place of origin.

It can only return to the redd where it was deposited as a fertilized egg if it knows where it is going and when it has arrived. Its objective is to reach a particular place in a particular stream within a particular time to deposit either its eggs or sperm, after which the salmon will die. But some of its offspring will live to run the same gauntlet of decisions when their time to spawn arrives.

Our lives have a common thread with that of the salmon, because every decision we make determines where we are, where we are going, and where we will end up. Our stream in life is the collective thinking of parental, peer, and social pressure. Like the salmon, which goes downstream with the current to the ocean, we often accept the route of least resistance of collective thinking embodied in the current world view.

Although most of the salmon die and become part of the sea, a few survive and begin swimming against the current to fulfill their life's purpose. As we mature, most of us will drown in the ocean of mass thinking, going with the current and seeking our sense of value from outside of ourselves through the acceptance of others who are also drowning in mass thinking.

A few, however, will chart their course against the current, driven by an inner need to find their life's fulfillment in and through sustainable community. They will dare to risk the unknown of continual change and fight their way upstream against the current of lackluster thinking embodied in the present world view. As they reach a place where materialism and spirituality are balanced, their focus will be so concentrated, their faith so strong, that what to others seems to be effort becomes to them increasingly effortless. They too, like the salmon, may die before sustainable community is achieved, but they will leave behind the seeds for even greater achievements by the next generation as it struggles toward social/environmental harmony.

No decision is easier or more difficult to make than another. The difficult part is getting ready to make the decision, which is a process of making many little, often unconscious, decisions to assess risk and benefit. Ultimately, one must decide how much criticism one is willing to take in swimming against the current to the purer waters beyond the nonsustainable, expansionist economic mentality. "I soon learned," said Albert Einstein, "to scent out the paths that led to the depths and to disregard everything else, all the many things that fill up the mind and divert it from the essential."

The more we dare to risk such criticism, however, the more focused we become, like the narrowing of the channel for the salmon. With increasing focus, the clarity of the vision we behold comes ever-closer to our grasp. This in turn brings clarity and decisiveness to our decisions and actions.

"Until one is committed," says Goethe, "there is hesitancy, the chance to draw back, always ineffectiveness. Concerning all acts of initiative, there is one elementary truth the ignorance of which kills countless ideas and splendid plans: that the moment one definitely commits oneself, then Providence moves too...whatever you can do or dream you can, begin it. Boldness has genius, power and magic in it. Begin it now."

"Sharp like the razor's edge, the sages say," through the Katha Upanishad, "is the path to self-realization." This is just another way of saying that we simply cannot get away from decisions. To avoid a decision is still to make a decision, although usually an unwise one. Nevertheless, we are not victims of life; we are products of our decisions. And our willingness to risk change dictates the boldness of our decisions.

We always make the best decision we can at a particular time, under a particular circumstance, with the data on hand. This does not mean that, given similar circumstances, we would make the same decision today or in the future. It only means that it was the best decision we could make at that time.

It does not mean that others will necessarily agree with our decisions or we with theirs. It only calls attention to the fact that I must accept your decision as your best because I cannot judge. I do not know why you did what you did; I only know what you did and how that appeared to me.

Besides, whether we realize it or not, whether we admit it or not, we need one another. Consider, for example, the large old trees of an ancient forest. Each signifies primeval majesty, but only together do they represent an ancient forest. Yet, we do not even see the forest for the trees.

If we could see belowground, we would find gossamer threads of a special kind of fungus stretching for billions of miles through the soil. These fungi grow as symbionts on and in the feeder roots of the ancient trees. Not only do they acquire food in the form of plant sugars through the roots of the ancient trees but also they provide nutrients, vitamins, and water from the soil to the trees and produce growth regulators that benefit the trees. These symbiotic fungus-root structures (called mycorrhizae) are the termini of the threads that form a complex fungal net under the entire ancient forest and, as evidence suggests, connect all trees one to another.

Like the ancient trees, we are separate individuals, and like the ancient forest united by its belowground fungi, we are united by our

humanity—our need for love, trust, respect, and unconditional acceptance of one another. As I look around the world, I see many won-
drous people in a great variety of sizes, shapes, and colors, each of
whom seems somehow separate from the rest.

I also see, however, that we must share our *feelings* with at least
one other person to find value in life. This tells me that when all is
said and done, we need one another because we, growing out of the
varied soils of culture, are united by the hidden threads of our common human needs. If, therefore, we lose sight of and touch with one
another as human beings, we will find a diminishing value in life. And
our common bonds will progressively erode into ever-increasing fear
and separateness.

Fear and separateness (which spawn destructive environmental
competition and political wars) is a choice made in secret in the
human heart and acted out in the collective of society. Love and
sustainable community (which foster trust, respect, and mutual caring)
is also a choice made in secret in the human heart and acted out in
the collective of society. And everyone has an equal choice, an equal
vote, if you will. With my choice, I influence the politics of life by how
I behave. With your choice, you do the same. Every choice counts,
like every vote in a democracy, and in the end, the majority will rule.

The parties are fear and love. The candidates are separateness (the
incumbent) and sustainable community (the challenger). How you
choose in the privacy of your own heart will determine how you
behave in public and will in turn influence the options we ensure
for the future, our collective legacy to the children of today and of
tomorrow.

ENDNOTES

1. Jay Walljasper. 1995. The second coming of the American city. *Resurgence* 172:42–43.

2. Gary Gardner. 1996. Preserving agricultural resources. pp. 78–94. *In*: Lester R. Brown, Janet Abramovitz, Chris Bright, et al. *State of the World 1996: A Worldwatch Institute Report on Progress Toward a Sustainable Society*. W.W. Norton, New York.

3. Jon Margolis. 1994. Theory of history suggests barbaric trends (Chicago Tribune). *Corvallis Gazette-Times* (Corvallis, OR) September 29.

4. Encarta. 1994. Microsoft Computer Encyclopedia.

5. Jay Griffiths. 1996. Life of strife. *Resurgence* 174:9–11.

6. F.F.H. Allen and Thomas W. Hoekstra. 1994. Toward a definition of sustainability. pp. 98–107. *In*: W. Wallace Covington and Leonard F. DeBano (Technical Coordinators). *Sustainable Ecological Systems: Implementing an Ecological Approach to Land Management*. USDA Forest Service General Technical Report RM-247, Rocky Mountain Forest and Range Experiment Station, U.S. Department of Agriculture, Fort Collins, CO.

7. John H. Baldwin. 1984. *Environmental Planning and Management*. Westview Press, Boulder, CO, 280 pp.; Dr. Kenneth S. Krane, Chair, Department of Physics, Oregon State University, Corvallis (personal communication).

8. David W. Orr. 1990. The question of management. *Conservation Biology* 4:8–9.

9. Nicholas Georgescu-Roegen. 1971. *The Entropy Law and the Economic Process*. Harvard University Press, Cambridge, MA.

10. Norman Jacob. 1989. Towards a theory of sustainability. *Trumpeter* 6:93–97.

11. Chris Maser. 1992. *Global Imperative: Harmonizing Culture and Nature.* Stillpoint Publishing, Walpole, NH, 267 pp.

12. Michael Kiefer. 1989. Fall of the Garden of Eden. *International Wildlife* July–August:38–43.

13. Fritz M. Heichelheim. 1956. The effects of classical antiquity on the land. pp. 165–182. *In*: W.L. Thomas (Ed.). *Man's Role in Changing the Face of the Earth.* University of Chicago Press, Chicago (in F.F.H. Allen and Thomas W. Hoekstra[6]).

14. Robert Cullen. 1993. The true cost of coal. *Atlantic Monthly* December:38, 40, 48–50, 51.

15. Walter C. Shortle and Ernest A. Bondietti. 1992. Timing, magnitude, and impact of acidic deposition on sensitive forest sites. *Water, Air, and Soil Pollution* 61:253–267.

16. R.J. Esher, D.H. Marx, S.J. Ursic, R.L. Baker, L.R. Brown, and D.C. Coleman. 1992. Simulated acid rain effects on fine roots, ectomycorrhizae, microorganisms, and invertebrates in pine forests of the southern United States. *Water, Air, and Soil Pollution* 61:269–278.

17. Charles E. Little. 1992. Report from Lucy's Woods. *American Forests* March/April:25–27, 68–69.

18. Paul Recer. 1995. Old pesticides spread across globe (The Associated Press). *Corvallis Gazette-Times* (Corvallis, OR) October.

19. The Associated Press. 1996. Study finds hidden menace behind beauty of rural Florida. *Corvallis Gazette-Times* (Corvallis, OR) January 21.

20. Elaine R. Ingham. 1995. Organisms in the soil: the functions of bacteria, fungi, protozoa, nematodes, and arthropods (Blue Mountains Natural Resources Institute, La Grande, OR). *Natural Resource News* 5:10–12, 16–17.

21. Chris Maser. 1995. The humble ditch. *Resurgence* 172:38–40.

22. Richard Plochmann. 1968. *Forestry in the Federal Republic of Germany.* Hill Family Foundation Series, School of Forestry, Oregon State University, Corvallis, 52 pp.

23. Richard Plochmann. 1989. The forests of Central Europe: a changing view. pp. 1–9. *In: Oregon's Forestry Outlook: An Uncertain Future.* The 1989 Starker Lectures. Forestry Research Laboratory, College of Forestry, Oregon State University, Corvallis.

24. John Bellamy Foster. 1995. Ecology and human freedom. *Monthly Review* 47:31.

25. Chris Maser. 1996. *Resolving Environmental Conflict: Toward Sustainable Community Development.* St. Lucie Press, Delray Beach, FL, 250 pp.

26. Duncan M. Taylor. 1992. Disagreeing on the basics. *Alternatives* 18: 26–33.

27. Jay Haley. 1994. Zen and the art of therapy. *Networker* January/February:55–60.

28. Margaret Shannon. 1992. Forest care: a feminist theory of forest management. pp. 72–92. *In: Culture and Natural Resources.* Starker Lectures. College of Forestry, Oregon State University, Corvallis.

29. Arnold Toynbee. 1958. *Civilization on Trial and the World and the West.* Meridian Books, New York, 348 pp.

30. Jeffery Mishlove. 1994. Intuition: the source of true knowing. *Noetic Sciences Review* 29:31–36.

31. Elizabeth Ann R. Bird. 1987. The social construction of nature: theoretical approaches to the history of environmental problems. *Environmental Review* 11:255–264.

32. Manfred Stanley. 1981. *Technological Conscience.* University of Chicago Press, Chicago.

33. Larry D. Harris. 1984. *The Fragmented Forest.* University of Chicago Press, Chicago, 211 pp.

34. Monica G. Turner. 1989. Landscape ecology: the effect of pattern on process. *Annual Review of Ecological Systems* 20:171–197.

35. Monica G. Turner, Eugene P. Odum, Robert Costanza, and Thomas M. Springer. 1988. Market and nonmarket values of the Georgia landscape. *Environmental Management* 12:209–217.

36. Kirk Talbott. 1993. *Central Africa's Forests: The Second Greatest Forest System on Earth.* World Resources Institute, Washington, D.C.

37. David Orr. 1995. Conservatives against conservation. *Resurgence* 172: 15–17.

38. Donella Meadows. 1995. Seven blunders. *Resurgence* 172:13.

39. Jean Houston. 1995. Deganawidah in the world. *The Quest* 8:10–17.

40. Robert Constanza and Herman E. Daly. 1987. Toward an ecological economics. *Ecological Modelling* 38:1–7.

41. Robert Rodale. 1988. Big new ideas—where are they today? Unpublished speech given at the Third National Science, Technology, Society (STS) Conference, February 5–7, Arlington, VA.

42. Max DePree. 1989. *Leadership Is an Art*. Dell Trade Paperback, New York, 148 pp.

43. C.J. George. 1972. The role of the Aswan Dam in changing fisheries of the south-western Mediterranean. *In*: M.T. Farvar and J.P. Milton (Eds.). *The Careless Technology*. Natural History Press, New York.

44. Sinn-Chye Ho. 1996. Vision 2020: towards an environmentally sound and sustainable development of freshwater resources in Malaysia. *GeoJournal* (in press).

45. Sandra Postel. 1996. Forging a sustainable water strategy. p. 59. *In*: Lester R. Brown, Janet Abramovitz, Chris Bright, et al. *State of the World 1996: A Worldwatch Institute Report on Progress Toward a Sustainable Society*. W.W. Norton, New York.

46. *The Holy Bible*, Authorized King James Version. World Bible Publishers, Iowa Falls, IA, Numbers 35:34.

47. Verne Gross Carter and Timothy Dale. 1974. *Topsoil and Civilization* (rev. ed.). University of Oklahoma Press, Norman.

48. W.C. Lowdermilk. 1975. *Conquest of the Land Through Seven Thousand Years*. Agricultural Information Bulletin No. 99. U.S. Department of Agriculture, Soil Conservation Service. U.S. Government Printing Office, Washington, D.C.

49. Ralph Metzner. 1995. Where is the first world? *Resurgence* 172:126–129.

50. Wendell Berry. 1990. Word and flesh. *Whole Earth Review* Spring:68–71.

51. Jeremy Rifkin. 1991. *Biosphere Politics: A Cultural Odyssey from the Middle Ages to the New Age*. Harper, San Francisco.

52. Ronald L. Warren. 1972. *The Community in America* (2nd ed.). Rand McNally College Publishing, Chicago.

53. Geoff Mulgan. 1995. A sense of community. *Resurgence* 172:18–20.

54. Anna F. Lemkow. 1994. Our common journey toward freedom. *The Quest* 7(1):55–63.

55. Kevin Phillips. 1994. *Arrogant Capital: Washington, Wall Street, and the Frustration of American Politics*. Little, Brown, New York, 231 pp.

56. The quote by Alexander Tayler was given to me by a friend who cannot remember where he got it, and I have found nothing on Alexander

Tayler in the literature. I have included the quote because it rings true, truer than the vast majority of Americans would probably care to admit.

57. Herman E. Daly and John V. Cobb. 1989. *For the Common Good: Redirecting the Economy Toward Community, Environment, and a Sustainable Future*. Beacon Press, Boston, 1,482 pp.

58. Gerald Gold and Richard Attenborough. 1983. The unfinished revolution. *Heart* Autumn:17–19, 108–112; Ramachandra Guha. 1995. Mahatma Gandhi and the environmental movement in India. *Capitalism, Nature, Socialism* 6:47–61.

59. Wendell Berry. 1995. Conserving communities. *Resurgence* 170:6–11.

60. Manly P. Hall. 1971. *Healing: The Divine Art*. Philosophical Research Society, Los Angeles, 341 pp.

61. M. Ravitz. 1982. Community development: challenge of the eighties. *Journal of the Community Development Society* 13:1–10.

62. The Associated Press. 1995. Experts call extinction rates "alarming." *Corvallis Gazette-Times* (Corvallis, OR) November 19.

63. Lester R. Brown. 1996. The acceleration of history. pp. 3–20. *In*: Lester R. Brown, Janet Abramovitz, Chris Bright, et al. *State of the World 1996: A Worldwatch Institute Report on Progress Toward a Sustainable Society*. W.W. Norton, New York.

64. D.W. Schindler, K.G. Beaty, E.J. Fee, D.R. Cruikshank, et al. 1990. Effects of climatic warming on lakes of the central boreal forest. *Science* 250:967–970.

65. Christopher Flavin. 1966. Facing up to the risks of climate change. pp. 21–39. *In*: Lester R. Brown, Janet Abramovitz, Chris Bright, et al. *State of the World 1996: A Worldwatch Institute Report on Progress Toward a Sustainable Society*. W.W. Norton, New York.

66. Luna B. Leopold. 1990. Ethos, equity, and the water resource. *Environment* 2:16–42.

67. T. Maddock, III, H. Banks, R. DeHan, R. Harris, J.H. Kneese, J.H. Lehr, P. McCarty, J. Mercer, D.W. Miller, M.L. Munts, M.A. Pierle, A.Z. Roisman, L. Swanson, and J.T.B. Tripp. 1984. *Protecting the Nation's Groundwater from Contamination*. OTA-0-233. Office of Technology Assessment, U.S. Congress, Washington, D.C., 244 pp.

68. D. Hand. 1990. Breadbasket ecology. *Yoga Journal* May/June 1990:23–24.

69. S. McCartney. 1986. Watering the west. Part 3. Growing demand, decreasing supply send costs soaring. *The Oregonian* (Portland, OR) September 30.

70. D.J. Chasan. 1977. *Up for Grabs, Inquiries into Who Wants What*. Madrona Publishers, Seattle, 133 pp.

71. W.J. Elliot, C.H. Luce, R.B. Foltz, and T.E. Koler. 1996. Hydrologic and sedimentation effects of open and closed roads (Blue Mountain Natural Resources Institute, La Grande, OR). *Natural Resource News* 6:7–8.

72. C.J. De Loach. 1971. The effect of habitat diversity on predation. *Proceedings Tall Timber Conference on Ecological Animal Control by Habitat Management* 2:223–241.

73. David Pimentel. 1971. Population control in crop systems: monocultures and plant spatial patterns. *Proceedings Tall Timber Conference on Ecological Animal Control by Habitat Management* 2:209–220.

74. N.W. Moore, M.D. Hooper, and B.N.K. Davis. 1967. Hedges. I. Introduction and reconnaissance studies. *Journal of Applied Ecology* 4:201–220.

75. Chris Maser. 1989. *Forest Primeval: The Natural History of an Ancient Forest*. Sierra Club Books, San Francisco, 282 pp.

76. Chris Maser. 1994. *Sustainable Forestry: Philosophy, Science, and Economics*. St. Lucie Press, Delray Beach, FL, 371 pp.

77. Chris Maser and James R. Sedell. 1994. *From the Forest to the Sea: The Ecology of Wood in Streams, Rivers, Estuaries, and Oceans*. St. Lucie Press, Delray Beach, FL, 200 pp.

78. Donald Ludwig, Ray Hilborn, and Carl Walters. 1993. Uncertainty, resource exploitation, and conservation: lesson from history. *Science* 260: 17, 36.

79. Noam Chomsky. 1995. How free is the free market? *Resurgence* 173:6–9.

80. Dave Broad. 1995. Globalization versus labor. *Monthly Review* 47(7):20–31; Peter Seybold. 1995. The politics of free trade: the global marketplace as a closet dictator. *Monthly Review* 47(7):43–48.

81. Frances Hutchinson. 1995. We are all economists. *Resurgence* 173:57.

82. Seth Shulman. 1995. Patent medicine. *Technology Review* 98:28–36.

83. Ashok Sharma. 1995. Poor countries want control over resources (The Associated Press). *Corvallis Gazette-Times* (Corvallis, OR) October 22.

84. Alan Ereira. 1995. Mayan medicine. *Resurgence* 173:59.

85. John Kinsman. 1995. Republican reforms threaten family farms (Knight-Ridder Tribune News Service). *Corvallis Gazette-Times* (Corvallis, OR) November 12.

86. Tracy Loew. 1995. Coast Range's private forest land controlled by a few, study says. *Corvallis Gazette-Times* (Corvallis, OR) November 9.

87. Steve Chase. 1995. Can capitalism be reformed? *The Trumpeter* 12:3–10.

88. Václav Havel. 1995. Democracy and transcendence. *Resurgence* 172:6–9.

89. Surur Hoda. 1995. Gandhi's talisman. *Resurgence* 172:10–13.

90. New Sunday Times. 1995. Seahorses become latest victim of overfishing. *The New Straits Times Press* (Kuala Lumpur, Malaysia) October 15.

91. The Associated Press. 1996. Should we save pandas and pigs? *Corvallis Gazette-Times* (Corvallis, OR) January 21.

92. Wendell Berry. 1992. Our tobacco problem. *Utne Reader* September/October:84–91.

93. Bernama. 1995. Centre: Asia to have water woes by 2025. *The Star* (Kuala Lumpur, Malaysia) October 11.

94. The Associated Press. 1995. U.N. report sounds alarm on dying species. *Corvallis Gazette-Times* (Corvallis, OR) November 14.

95. James Freed. 1995. Special forest products: past, present, future. *International Journal of Ecoforestry* 11:62–67.

96. Brother David Steindl-Rast. 1984. *Gratefulness and the Heart of Prayer: An Approach to Life in Fullness*. Paulist Press, Ransey, NJ, 224 pp.

97. Elizabeth Brown. 1995. Voluntary simplicity (Journal American). *Corvallis Gazette-Times* (Corvallis, OR) June 20.

98. Diane Lewis. 1995. In search of the simple life (Knight-Ridder Tribune News Service). *Corvallis Gazette-Times* (Corvallis, OR) November 5.

99. Elaine St. James. *Simplify Your Life: 100 Ways to Slow Down and Enjoy the Things That Really Matter*. Hyperion, New York ($7.95); Andy Dappen. *Cheap Tricks: 100s of Ways You Can Save 1000s of Dollars*. Brier Books, Box 180, Mountlake Terrace, WA 98043, (800) 742-4847 ($15); Vicki Robin and Joe Dominguez. *Your Money or Your Life*. Random House, New York ($11.95); Duane Elgin. *Voluntary Simplicity: Toward a Way of Life That Is Outwardly Simple, Inwardly Rich*. William Morrow, New York ($10); *Simple Living*. 2319 N. 45th, Box 149, Seattle, WA 98103, (206) 464-4800 ($14 per year, quarterly); *Tightwad Gazette*. R.R. 1, Box 3570 Leeds, ME 04263 ($12 per year, monthly).

100. Karl-Henrik Robert. 1991. Educating a nation: the natural step. *In Context* 28:10–15.

101. Mark S. Meisner. 1995. Metaphors of nature: old vinegar in new bottles? *The Trumpeter* 12:11–18.

102. Per Bak and Kan Chen. 1991. Self-organizing criticality. *Scientific American* 267:46–53.

103. William K. Stevens. 1990. New eye on nature: the real constant is eternal turmoil. *The New York Times* July 31.

104. John J. Magnuson. 1990. Long-term ecological research and the invisible present. *BioScience* 40:495–501.

105. The Associated Press. 1995. Willamette study shows deformed fish. *Corvallis Gazette-Times* (Corvallis, OR) October 24.

106. Peter Harper. 1995. Natural sewage systems. *Resurgence* 173:28–29.

107. Manfred Stanley. 1983. The mystery of the commons: on the indispensability of civil rhetoric. *Social Research* 50:851–883.

108. Mary Douglas. 1986. *How Institutions Think.* Syracuse University Press, Syracuse, NY.

109. Vinoba Bhave. 1994. Moved by love. *Resurgence* 165:26–27.

110. Cecile Andrews. 1992. Study circles: schools for life. *In Context* 33:22–25.

111. Gifford Pinchot and Elizabeth Pinchot. 1994. Beyond bureaucracy. *Business Ethics* 8(2).

112. Elizabeth Pinchot and Gifford Pinchot. 1994. *The End of Bureaucracy and the Rise of the Intelligent Organization.* Berrett-Koehler, San Francisco; Gifford Pinchot and Elizabeth Pinchot. 1993. Unleashing intelligence. *Executive Excellence* September:7–8; Elizabeth Pinchot. 1992. Can we afford ethics? *Executive Excellence* March:1–2; Elizabeth S. Pinchot. 1992. Balance the power. *Executive Excellence* September:3–4; Gifford Pinchot. 1992. Rewarding with status. *Executive Excellence* August:3–5.

113. Austine Roberts. 1977. *Centered on Christ: An Introduction to Monastic Profession.* St. Bede's Publications, Still Rive, MA, 169 pp.

114. Joseph Campbell. 1968. *The Hero with a Thousand Faces.* Bollingen Series, Princeton University Press, Princeton, NJ, 416 pp.

115. *The Holy Bible,* Authorized King James Version. World Bible Publishers, Iowa Falls, IA, Philippians 4:8.

REFERENCES

Allen, F.F.H and Thomas W. Hoekstra. 1994. Toward a definition of sustainability. pp. 98–107. *In*: W. Wallace Covington and Leonard F. DeBano (Technical Coordinators). *Sustainable Ecological Systems: Implementing an Ecological Approach to Land Management*. USDA Forest Service General Technical Report RM-247. Rocky Mountain Forest and Range Experiment Station, U.S. Department of Agriculture, Fort Collins, CO.

Bak, P. and K. Chen. 1991. Self-organizing criticality. *Scientific American* January:46–53.

Baumhoff, M.A. and R.F. Heize. 1967. Postglacial climate and archaeology in the desert west. pp. 697–707. *In*: J.E. Wright, Jr. and D.G. Frey (Eds.). *The Quaternary of the United States*. Princeton University Press, Princeton, NJ.

Brown, Lester R., Janet Abramovitz, Chris Bright, et al. 1996. *State of the World 1996: A Worldwatch Institute Report on Progress Toward a Sustainable Society*. W.W. Norton, New York, 249 pp.

Campbell, J. 1988. *The Power of Myth*. Doubleday, New York, 233 pp.

Covington, W.W. and M.M. Moore. 1991. Changes in forest conditions settlement. Unpublished report, submitted to J. Keane, Water Resources Operations, Salt River Project, Phoenix, AZ, 50 pp.

Davis, Margaret B. 1989. Lags in vegetation response to greenhouse warming. *Climatic Change* 15:75–82.

Davis, Margaret B. and C. Zabinski. 1991. Changes in geographical range resulting from greenhouse warming effects on biodiversity in forests. *In:* R.L. Peters and T.E. Lovejoy (Eds.). *Consequences of Global Warming for Biodiversity: Proceedings of the World Wildlife Fund Conference*. Yale University Press, New Haven, CT.

Delcourt, Hazel R. and Paul A. Delcourt. 1988. Quaternary landscape ecology: relevant scales in space and time. *Landscape Ecology* 2:23–44.

Delcourt, Paul A. and Hazel R. Delcourt. 1985. Dynamic landscapes of east Tennessee: an integration of paleoecology, geomorphology, and archaeology. *Studies in Geology* 9:191–220.

de Steiguer, J.E., J.M. Pye, and C.S. Love. 1990. Air pollution damage to U.S. forests. *Journal of Forestry* 88:17–22.

Dillon, L.S. 1956. Wisconsin climate and life zones in North America. *Science* 123:167–176.

Dix, R.L. 1964. A history of biotic and climatic changes within the North American grassland. pp. 71–89. *In:* D.J. Crisp (Ed.). *Grazing in Terrestrial and Marine Environments*. Blackwell Science Publishing, England.

Dobson, Andy, Alison Jolly, and Dan Rubenstein. 1989. The greenhouse effect and biological diversity. *Tree* 4:64–68.

Dorf, E. 1960. Climatic changes of the past and present. *American Scientist* 48:341–346.

Drengson, Alan R. 1985. The virtue of Socratic ignorance. pp. 34– 42. *In:* J.C. Edwards and D.M. MacDonald (Eds.). *Occasions for Philosophy*. Prentice-Hall, Englewood Cliffs, NJ.

Durbin, Kathie. 1991. Marbled murrelet may join threatened species. *The Oregonian* (Portland, OR) June 18.

Edgar, C. and B. Adams. 1992. *Ecology and Decline of Red Spruce in the Eastern United States*. Springer-Verlag, New York.

Esher, R.J., D.H. Marx, S.J. Ursic, R.L. Baker, L.R. Brown, and D.C. Coleman. 1992. Simulated acid rain effects on fine roots, ectomycorrhizae, microorganisms, and invertebrates in pine forests of the southern United States. *Water, Air, and Soil Pollution* 61:269–278.

Franklin, J.F. and R.T.T. Forman. 1987. Creating landscape patterns by forest cutting: ecological consequences and principles. *Landscape Ecology* 1:5–18.

George, C.J. 1972. The role of the Aswan Dam in changing fisheries of the south-western Mediterranean. *In:* M.T. Farvar and J.P. Milton (Eds.). *The Careless Technology*. Natural History Press, New York.

Graham, A. and C. Heimsch. 1960. Pollen studies of some Texas peat deposits. *Ecology* 41:751–763.

Griffin, J.B. 1967. Late Quaternary prehistory in the northeastern woodlands. pp. 655–667. *In:* J.E. Wright, Jr. and D.G. Frey (Eds.). *The Quaternary of the United States*. Princeton University Press, Princeton, NJ.

Guilday, J.E., P.W. Parmalee, and H.W. Hamilton. 1977. The Clark's Cave bone deposits and the late Pleistocene paleoecology of the Central Appalachian Mountains of Virginia. *Carnegie Museum of Natural History Bulletin* 2:1–87.

Hardin, G. 1984. *An Ecolate View of the Human Predicament.* The Environmental Fund, Monograph Series, 14 pp.

Hardin, G. 1986. Cultural carrying capacity: a biological approach to human problems. *BioScience* 36:599–606.

Harris, L.D. 1984. *The Fragmented Forest.* University of Chicago Press, Chicago, 211 pp.

Harris, L.D. and C. Maser. 1984. Animal community characteristics. pp. 44–68. *In*: L.D. Harris. *The Fragmented Forest.* University of Chicago Press, Chicago.

Ho, S-C. 1994. Status of limnological research and training in Malaysia. *Mitt. Internat. Verein. Limnol.* 24:129–145.

Ho, S-C. 1996. Vision 2020: towards an environmentally sound and sustainable development of freshwater resources in Malaysia. *GeoJournal* (in press).

Hopkins, D.M. 1959. Cenozoic history of the Bering Land Bridge. *Science* 129:1519–1528.

Jacob, N. 1989. Towards a theory of sustainability. *Trumpeter* 6:93–97.

Lélé, Sharachandra M. 1991. Sustainable development: a critical review. *World Development* 19:607–621.

Maser, Chris. 1992. *Global Imperative: Harmonizing Culture and Nature.* Stillpoint Publishing, Walpole, NH, 267 pp.

Maser, Chris. 1995. *Resolving Environmental Conflict: Towards Sustainable Community Development.* St. Lucie Press, Delray Beach, FL, 250 pp.

Monastersky, R. 1991. Hot year prompts greenhouse concern. *Science News* 139:36.

Montgomery, Claire A. and Robert A. Pollak. 1995. Economics and biodiversity. *Journal of Forestry* (in press).

Montgomery, Claire A. and Robert A. Pollak. 1995. Valuing and measuring biodiversity for comparing land-use alternatives. pp. 1–8. *In*: Proceedings of the IUFRO XX World Congress, August 6–12, 1995, Tempere, Finland.

Morrison, Peter H. and Fredrick J. Swanson. 1990. *Fire History and Pattern in a Cascade Range Landscape.* USDA Forest Service General Technical Report PNW-GTR-254. Pacific Northwest Research Station, Portland, OR, 77 pp.

Myers, N. and R. Tucker. 1987. Deforestation in Central America: Spanish legacy and North American consumers. *Environmental Review* 11:55–71.

Nault, L.R. and W.R. Findley. 1981. Primitive relative offers new traits to improve corn. *Ohio Report* 66 (Nov./Dec.):90–92.

Olson R.K. and A.S. Lefohn (Eds.). 1989. *Transactions Effects of Air Pollution on Western Forests*. Air and Waste Management Association, Pittsburgh, PA, 577 pp.

Perry, David A. 1988. An overview of sustainable forestry. *Journal of Pesticide Reform* 8:8–12.

Perry, David A. 1988. Landscape pattern and forest pests. *Northwest Environmental Journal* 4:213–228.

Perry, David A. and Jeffrey G. Borchers. 1990. Climate change and ecosystem responses. *Northwest Environmental Journal* 6:293–313.

Perry, D.A., M.P. Amaranthus, J.G. Borchers, S.L. Borchers, and R.E. Brainerd. 1989. Bootstrapping in ecosystems. *BioScience* 39:230–237.

Perry, D.A., J.G. Borchers, S.L. Borchers, and M.P. Amaranthus. 1990. Species migrations and ecosystem stability during climate change: the belowground connection. *Conservation Biology* 4:266–274.

Péwé, T.L., D.M. Hopkins, and J.L. Giddings. 1967. The Quaternary geology and archaeology of Alaska. pp. 355–374. *In:* J.E. Wright, Jr. and D.G. Frey (Eds.). *The Quaternary of the United States.* Princeton University Press, Princeton, NJ.

Phillips, Kevin. 1994. *Arrogant Capital: Washington, Wall Street, and the Frustration of American Politics.* Little, Brown, New York, 231 pp.

Rapport, David J. 1989. What constitutes ecosystem health? *Perspectives in Biology and Medicine* 33:120–132.

Rapport, D.J, H.A. Regier, and T.C. Hutchinson. 1985. Ecosystem behavior under stress. *The American Naturalist* 125:617–640.

Schlesinger, M.E. and F.F. Mitchell. 1985. Model predictions of the equilibrium climatic response to increased carbon dioxide. pp. 83–147. *In: Projecting the Climatic Effects of Increasing Carbon Dioxide.* DOE/ER-0237. U.S. Department of Energy, Washington, D.C.

Schowalter, T.D. 1985. Adaptations of insects to disturbance. pp. 235–386. *In:* S.T.A. Pickett and P.S. White (Eds.). *The Ecology of Natural Disturbance and Patch Dynamics.* Academic Press, New York.

Schowalter, T.D. 1988. Forest pest management: a synopsis. *Northwest Environmental Journal* 4:313–318.

Schowalter, T.D. 1989. Canopy arthropod community structure and herbivory in old-growth and regenerating forests in western Oregon. *Canadian Journal of Forestry Research* 19:318–322.

Schowalter, T.D and J.E. Means. 1988. Pest response to simplification of forest landscapes. *Northwest Environmental Journal* 4:342–343.

Schowalter, T.D and J.E. Means. 1989. Pests link site productivity to the landscape. pp. 248–250. *In*: D.A. Perry, R. Meurisse, B. Thomas, R. Miller, et al. (Eds.). *Maintaining the Long-Term Productivity of Pacific Northwest Forest Ecosystems*. Timber Press, Portland, OR.

Schowalter, T.D., W.W. Hargrove, and D.A. Crossley, Jr. 1986. Herbivory in forested ecosystems. *Annual Review of Entomology* 31:177–196.

Shearman, R. 1990. The meaning and ethics of sustainability. *Environmental Management* 14:108.

Slobodkin, Lawrence B. 1988. On the susceptibility of different species to extinction: elementary instruction for owners of a world. pp. 226–242. *In*: B.G. Norton (Ed.). *The Preservation of Species*. Princeton University Press, Princeton, NJ.

Steiner, Rudolf. 1919. *The Threefold Social Order* (translation from *The Renewal of the Social Organism*. 1985). The Anthroposophic Press, Hudson, NY.

Stephenson, R.L. 1967. Quaternary human occupation of the plains. pp. 685–696. *In*: J.E. Wright, Jr. and D.G. Frey (Eds.). *The Quaternary of the United States*. Princeton University Press, Princeton, NJ.

Swetnam, Thomas W. 1988. Forest fire primeval. *Natural Science* 3:236–241.

Swetnam, Thomas W. 1990. Fire history and climate in the southwestern United States. pp. 6–17. *In*: J.S. Krammers (Tech. Coord.). *Effects of Fire in Management of Southwestern Natural Resources*. USDA Forest Service General Technical Report RM-191. Rocky Mountain Research Station, Fort Collins, CO.

Turner, Monica G. 1989. Landscape ecology: the effect of pattern on process. *Annual Review of Ecological Systems* 20:171–197.

World Commission on Environment and Development. 1987. *Our Common Future*. Oxford University Press, New York.